Springer Praxis Books

Artist's impression

More information about this series at http://www.springer.com/series/4097

Fabio Vittorio De Blasio

Mysteries of Mars

Fabio Vittorio De Blasio
Earth and Environmental Sciences
University of Milan Bicocca
Milan, Milano, Italy

Springer Praxis Books
Popular Astronomy
ISBN 978-3-319-74783-5 ISBN 978-3-319-74784-2 (eBook)
https://doi.org/10.1007/978-3-319-74784-2

Library of Congress Control Number: 2018940009

This Praxis imprint is published by the registered company Springer Nature Switzerland AG
The registered company address is: Gewerbestrasse 11, 6330 Cham, Switzerland

Per aspera ad Astra

Topography of Mars

Preface

Few fields of science are as interesting as the exploration of the solar system. New discoveries are made in series when a space mission reaches a target and newspapers feature enigmatic images of distant worlds. Even if the novelty fades, diluted in a sea of terrestrial matters now perceived as more urgent but soon forgotten, the technical and scientific knowledge gained by humanity from the interplanetary missions accumulates permanently. And also, between one mission and the next, when the public's interest decreases, novelties appear almost every day: theories, models, and interpretations about the origin and evolution of the solar system. But the king of the planets remains Mars—akin to Earth, yet so different. Devoid of abundant life as we know it here on Earth, but not of tornadoes and storms, huge volcanoes, catastrophic landslides, breathtaking views, and once, a long time ago, also of glaciers, rivers, lakes, and perhaps entire oceans. It is a planet that we can explore with great precision: the only one on which man will set foot. The increase of knowledge about Mars has been explosive rather than linear: comparing Martian maps of the 1960s with current images, it is hard to see much similarity. The planet of half a century ago was a collage of weak spots, doubtful channels, nonexistent details, and blurred white areas. Everything changed after the Mariner 4 mission, the first to provide details of the Martian surface. When the spacecraft reached the planet in 1965, scientists held their breath. They anticipated lands separated by oceans, canals, and perhaps even plant life. However, Mars seemed at first a rather disappointing planet, barren like the Moon. But subsequent missions overturned this sterile picture, and under the lens of increasingly sophisticated equipment, Mars began to show its very active past and enigmatic side. While some geological structures, such as the huge impact basins, had no equal on Earth, patterns and torrential outflow channels were similar to those on our planet. Yet, judging by the dry climate of today, Mars should not have had rivers and streams, unless the climate has changed completely over the course of its history!

This book introduces the reader to the wonders of Mars. Photos and data from half a century of missions are still being processed. We know a great deal about our planetary cousin, but many aspects of its history remain enigmatic. Ten major mysteries are introduced in parallel. Where has the water gone, once so abundant on the red planet? How do we know that the climate of Mars has changed over billions of years? And why does Mars feature two distinct parts, a northern hemisphere dominated by lowlands and a mountainous south? Shedding light on the mysteries of Mars is not only inspiring for those interested in astronomy and planetary science. It is also a keystone for understanding the Earth itself and our place in the cosmos. As regards the exploration of Mars and the planets, we are fortunate to live in a period of such great discoveries, like the one that must have animated the explorers of the ocean in the sixteenth century. Our exploration is only conceptual, without any physical risk. But it is by no means less exciting.

Tips for Reading

Of all the bodies that orbit the Sun and their satellites, Mars is the most similar to the Earth. This is why the book makes great use of terrestrial analogies, with unusual images for a book of this kind. If some of the morphologies of Mars can be understood on the basis of similar terrestrial structures past or present, others appear to be peculiar to Mars. The lack of a terrestrial reference makes them unique and enigmatic.

This book makes no claim of completeness. It is not a textbook, nor will it make the reader erudite on this subject. The hope is to stimulate curiosity about this still so mysterious planet. Many topics initially planned, such as the interior layered deposits, Martian moons, and the lakes of Mars, were eventually excluded due to lack of space. Others, such as remote sensing mineralogical analyses or rover data, especially Curiosity, were excluded from the beginning, as they are covered by other recent books. The emphasis is, above all, on the geomorphology of Mars, as deduced from optical and infrared images. There is also a didactic intent, an invitation to high school students and students of the first three years of university to undertake the study of planetary sciences. This is evident not only from the attempt to make the topic even more interesting by emphasizing the holes in our knowledge, but also in a series of "technical boxes" in which there are also practical indications (necessarily short) on how one can experiment a bit with the images coming from the main spacecraft. I wonder if any student or enthusiast can perhaps find something new!

I have been unconcerned with the references. Not wishing to pack the book with references (this is not a professional book), I indicate as footnotes only those references in which one author proposes a model different from the standard interpretation of some structure. Otherwise, when something is "common wisdom" and quite uncontroversial, I have not bothered about referencing the first scientists who proposed that interpretation. "FVDB" in a caption means image created by the author. "mf" means modified by the author.

Milan, Italy Fabio Vittorio De Blasio

Contents

Contents of Martian Mysteries

Technical Boxes

Chapter 1
Mars Through the Millennia

For the Egyptians it was "the red one," for the Sumerians, the star of death Nergal. It was Ares to the Greeks and Mars to the Romans. For most, it represented the god of war. It must have been the red color of the planet, evoking blood or the eyes of a furious animal, that suggested a violent personality for Mars. In the Middle Ages, being born with Mars in one's zodiac sign could make you violent and vicious. But it was thanks to Mars that Kepler came to understand the laws of planetary motion, paving the way for the theory of universal gravitation of Newton.On the following pages, we will consider the myths about Mars built by ancient cultures, but also their valuable observations; the first attempts to observe the surface of the planet with telescopes; how some remarkable discoveries were made, while other features induced early observers to ruminate about huge engineering feats built by aliens. We will see how our understanding of the Red Planet changed with the observations made in space, and familiarize ourselves with some of the characteristics of the planet.

F. V. De Blasio, *Mysteries of Mars*, Springer Praxis Books,
https://doi.org/10.1007/978-3-319-74784-2_1

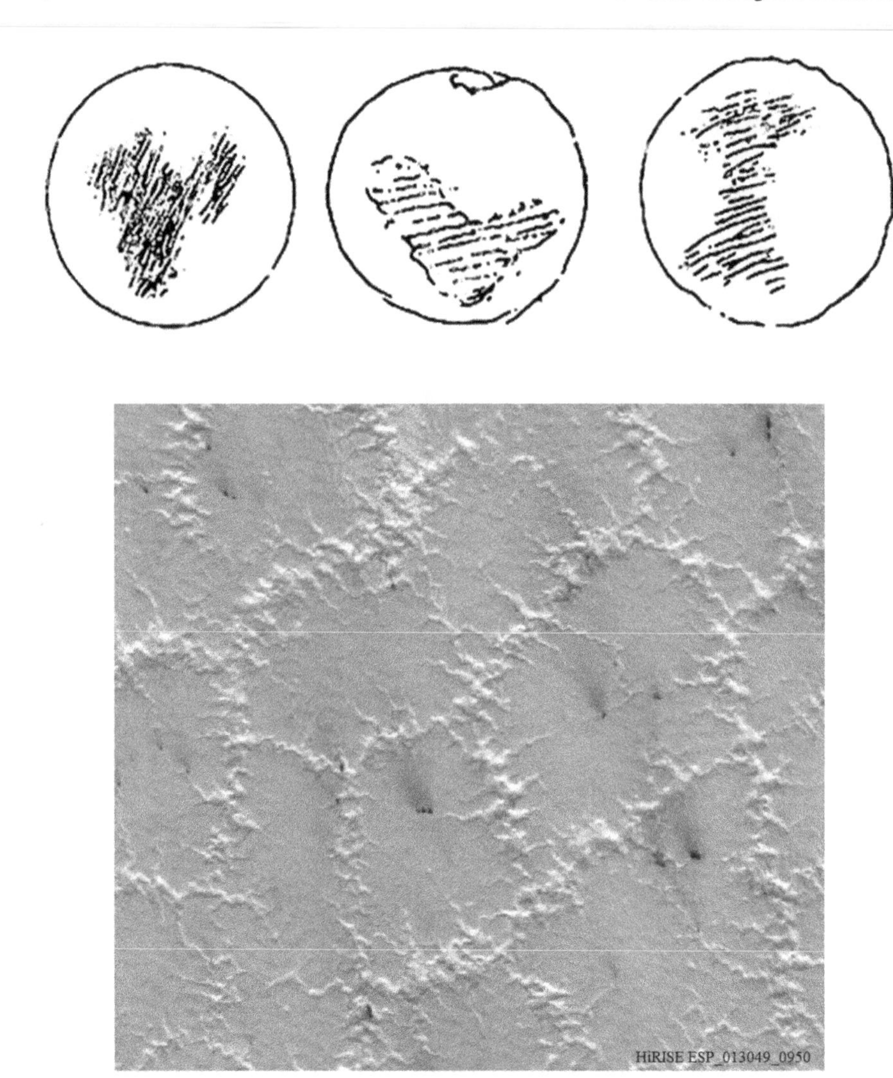

The above figure shows a comparison between the first published map of Mars (Huygens 1696) and a detailed view of the surface taken in 2009 by the high-resolution camera HiRISE. The latter portrays enigmatic structures in dry south polar ice called "spiders." The black spots are perhaps the outlet holes of jets of carbon dioxide similar to terrestrial geysers. Three centuries separate the images drawn by Huygens from the modern one, in which details less than one meter across can, in principle, be appreciated. Public Domain and HiRISE (MRO, NASA)

1.1 Approaching Mars

July 4, 1965, American Independence day. After decades of theories, centuries of studies, and millennia of myths, the basic questions about Mars and the possible existence of Martians are finally heading toward an answer. The spaceship Mariner 4, launched by NASA 7 months earlier, has finally reached the best position for the first close-up pictures of the Red Planet. Only 22 images; this is what the technology of the time permits. The photos are first recorded on magnetic tape and then transmitted at a rate of only 8 bits per second, millions of times slower than the transmission of a modern electronic message.

In the decades preceding Mariner 4, scientists and philosophers had speculated about the existence of life on Mars, perhaps evolved and intelligent. How else could one explain these seemingly artificial channels, tens of kilometers wide and thousands of kilometers long? Many people were utterly convinced of the existence of Martians and some engineers had designed bizarre methods to communicate with them; pessimists even feared an invasion from outer space. In 1938, a young Orson Welles broadcast a radio show in America. The *War of the Worlds* was a transposition of a book written by the English author Herbert George Wells 40 years earlier. In the story, aggressive Martians invade the Earth, using Victorian England as a bridgehead. For some days, many Americans believed uncritically that an alien invasion was in progress. Panic spread, in spite of the announcement that the radio play was pure fantasy.

Many scientists, though not very convinced of the existence of little green men, hoped at the very least to find Earth's sister planet in Mars. The maps drawn through the telescope by Slipher, the most recent and reliable, showed channels separated by unknown lands; perhaps not dug by Martians, but full of water anyway.

At the available rate of transmission, it takes time for all of the bits of information from Mariner 4 to reach the Earth. This is not the only problem: the signal strength decreases a billion times in the ten-light-minute-long journey. Those subtle pulses, billionths of a watt in power, must be transformed into something intelligible. When, after 9 days, nineteen useful frames have finally arrived (three having been lost), the scientists at NASA have gained an impression of a world very different from the idyllic place that was expected. Something very surprising.

Mars as a Heavenly Warrior

In all human cultures, the planets were considered the most important celestial objects after the Sun and the Moon. The reason is simple: while the stars shine in the same place every night (except for the very slow phenomenon of the precession of the equinoxes), the planets travel in the sky. They visit different constellations over the years; how could we humans not attribute a value to these mysterious cosmic movements?

Fig. 1.1 Left: According to the astrology of the Middle Ages, those who are born with Mars in their zodiacal sign might become soldiers, but also murderers. Right: the Mayans had an obsession with the planets, especially Venus. Here, the glyph representing Mars is shown. Left: public domain, source unknown, right: FVDB

The ancients attributed at least three meanings to the planets. The first was mythological: each planet was a divine personage who, for some (usually earthly) reason, ended up in the cosmos. For the Sumerians, it was Nergal (or Nirgal), the god of war, just like the Greek Ares and Mars of the Romans. Among the duties of the Salii, priests of ancient Rome consecrated to Mars, was opening the military year in March and closing it in September. Special ceremonies focused on dances inspired by the art of war. It must have been the red color of the planet, recalling blood or the eyes of a furious animal, that suggested a violent role for Mars.

The second significance was astrological. In an era of magical thinking, the movement of the planets through the different constellations ensured a symbolism vital for predicting the future (something that is believed by many people even today). In the Middle Ages, to be born with Mars in one's zodiac sign signified a future of violence and thievery (Fig. 1.1). Even the value of the geometric and scientific planetary motion was immediately recognized by various civilizations. Early Chaldean and Greek observers wondered why those bright dots were travelling across the sky, always following an imaginary line that the Sun also tracks throughout the year, the ecliptic. It was soon realized that Mars had a synodic period of 780 days, after which its position in the sky was back to where it began. Some people

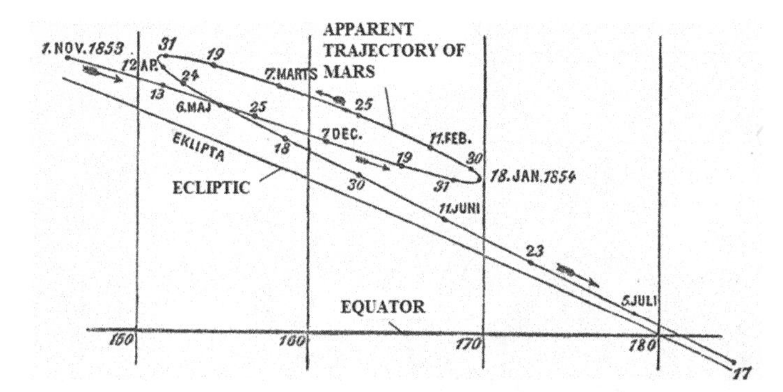

Fig. 1.2 The trajectory of Mars on the celestial sphere during the days between November 1, 1853 and July 17, 1854. Notice the loop that Mars makes between November 1853 and May 1854, making it move retrograde from January 18, 1854 to March 31. P. La Cour and J. Appel, Historisk Fysik, 1906

believed these cycles were of cosmic importance. This was the case for pre-Columbian civilizations, whose agricultural activities, military and civil decisions, as well as human and animal sacrifices, were established in harmony with the movement of stars and planets. There are reports of special Mayan rites based on rare planetary conjunctions. The Dresden Codex contains many numbers that are multiples of 78, perhaps a reference to the synodic period of Mars (Fig. 1.1). Lacking the ability to understand that the planets were spherical worlds like Earth, the ancients did what they could do: they measured their positions in the sky with precision. And in doing so, they found something inexplicable.

Mars, with Its Odd Orbit, Enters Science

Early observers noticed that Mars and the other known planets did not travel at a uniform rate along the ecliptic. Normally, they would travel along the same direction as that of the Sun and the Moon (prograde motion), but sometimes, they reversed direction (retrograde motion). As a consequence of such inversions, the planets' trajectories drew strange loops on the celestial sphere. One of these loops is shown in Fig. 1.2; it depicts the positions of Mars in the sky from November 1, 1853 to July 5 of the following year. Ptolemy had understood that such a complex trajectory would result from the combination of at least two different movements. In his conception, a planet revolved around a small circle called the epicycle, whose center was, in turn, carried by a much larger circle, the deferent.

The Ptolemaic model was conceptually complicated but could explain the observations, because they were constructed exactly for that purpose. Today, we know that the strange prograde and retrograde movement is due to a combination of the motion of Mars with that of the Earth. The need to abandon the complex model of epicycles and deferent cycles led to the Copernican model of the universe.

Johannes Kepler.

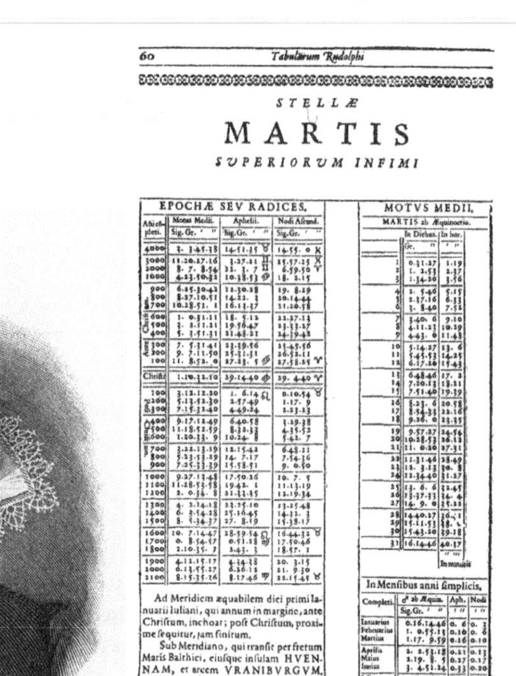

60 Tabularum Rudolphi

STELLÆ

MARTIS

SVPERIORVM INFIMI

EPOCHÆ SEV RADICES.

Anni completi.	Motus Medii. Sig. Gr. ′ ″	Aphelii. Sig. Gr. ′ ″	Nodi Ascend. Sig. Gr. ′ ″
4000	3. 3.45.38	14.51.35 ♉	14.55. 0 ♓
3000	11.20.27.16	3.27.21 ♊	25.57.25 ♓
2000	8. 7. 8.54	22. 3. 7 ♊	6.59.50 ♈
1000	4.23.50.32	10.38.53 ♋	18. 2.15
900	6.25.30.42	12.30.38	19. 8.29
800	8.27.10.51	14.22. 3	20.14.44
700	10.28.51. 1	16.13.37	21.20.58
600	1. 0.31.11	18. 5.12	22.27.13
500	3. 2.11.21	19.56.47	23.33.27
400	5. 3.51.31	21.48.21	24.39.42
300	7. 5.31.41	23.39.56	25.45.56
200	9. 7.11.50	25.31.31	26.52.11
100	11. 8.52. 0	27.23. 5 ♋	27.58.25 ♈
Christi	1.10.32.10	29.14.40 ♋	29. 4.40 ♈
100	3.12.12.20	1. 6.14 ♌	0.10.54 ♉
200	5.13.52.30	2.57.49	1.17. 9
300	7.15.32.40	4.49.24	2.23.23
400	9.17.12.49	6.40.58	3.29.38
500	11.18.52.59	8.32.33	4.35.52
600	1.20.33. 9	10.24. 8	5.42. 7
700	3.22.13.19	12.15.42	6.48.21
800	5.23.53.29	14. 7.17	7.54.36
900	7.25.33.39	15.58.51	9. 0.50
1000	9.27.13.48	17.50.26	10. 7. 5
1100	11.28.53.58	19.42. 1	11.13.19
1200	2. 0.34. 8	21.33.35	12.19.34
1300	4. 2.14.18	23.25.10	13.25.48
1400	6. 3.54.28	25.16.45	14.32. 3
1500	8. 5.34.37	27. 8.19	15.38.17
1600	10. 7.14.47	28.59.54 ♌	16.44.32 ♉
1700	0. 8.54.57	0.51.28 ♍	17.50.46
1800	2.10.35. 7	2.43. 3	18.57. 1
1900	4.12.15.17	4.34.38	20. 3.15
2000	6.13.55.27	6.26.12	21. 9.30
2100	8.15.35.36	8.17.46 ♍	22.15.45 ♉

Ad Meridiem æquabilem diei primi Ianuarii Iuliani, qui annum in margine, ante Christum, inchoat; post Christum, proxime sequitur, jam finitum.

Sub Meridiano, qui transit per fretum Maris Balthici, eiusque insulam HVENNAM, et arcem VRANIBVRGVM.

Ante Christum Anno 3993. die 24. Iulii, Vraniburgi H. 0. 33′. 26″.

	Medius ♂	Aphelium ♂	Nodus asc. ♂
	10 43′. 52″ ♋	15. 0′. 0 ♉	15. 0′. 0″ ♓
Quidsi	0. 0′. 0″ ♋	0. 0. 0 ♈ vel ♋	0.0.0 ♈

MOTVS MEDII.

MARTIS ab Æquinoctio.

	In Diebus. Gr. ′ ″	In hor. ′ ″
1	0.31.27	1.19
2	1. 2.53	2.37
3	1.34.20	3.56
4	2. 5.46	5.15
5	2.37.16	6.33
6	3. 8.40	7.52
7	3.40. 6	9.10
8	4.11.33	10.29
9	4.43. 0	11.48
10	5.14.27	13. 6
11	5.45.53	14.25
12	6.17.20	15.43
13	6.48.46	17. 2
14	7.20.13	18.21
15	7.51.40	19.39
16	8.23. 6	20.58
17	8.54.33	22.16
18	9.26. 0	23.35
19	9.57.27	24.54
20	10.28.53	26.13
21	11. 0.20	27.31
22	11.31.46	28.49
23	12. 3.13	30. 8
24	12.34.40	31.27
25	13. 6. 6	32.45
26	13.37.33	34. 4
27	14. 9. 0	35.22
28	14.40.27	[illegible]
29	15.11.53	[illegible]
30	15.43.20	39.18
31	16.14.46	40.37
		″ ‴ In minutis

In Mensibus anni simplicis,

Completi.	♂ ab Æquin. Sig. Gr. ′ ″	Aph. ′ ″	Nodi ′ ″
Ianuarius	0.16.14.46	0. 6	0. 3
Februarius	1. 0.55.13	0.10	0. 6
Martius	1.17. 9.59	0.16	0.10
Aprilis	2. 2.53.18	0.21	0.13
Maius	2.19. 8. 5	0.27	0.17
Iunius	3. 4.51.24	0.33	0.20
Iulius	3.21. 6.11	0.38	0.23
Augustus	4. 7.20.57	0.43	0.27
September	4.23. 4.16	0.49	0.30
October	5. 9.19. 3	0.55	0.34
November	5.25. 2.22	1. 1	0.37
December	6.11.17. 8	1. 7	0.40

MOTVS

Fig. 1.3 Left: Johannes Kepler utilized the observations of Tycho Brahe to the greatest extent, especially those on Mars, and ultimately deduced the laws of planetary motion. Right: Tablets of the motion of Mars from Tabulae Rudulphinae (1627). Left: FOTOLIA image 37564455/nickolae; right: public domain, source unknown

It was Johannes Kepler (1571–1630, Fig. 1.3) who conceived of Mars being one of the most important tools in the study of the laws of planetary motion, from which, in 1665, Newton began to establish the law of gravitation. Published in 1609, *De Martis motibus stellae* announced Kepler's first two laws: the orbits of the planets are ellipses and not circles, with the Sun at one focus. The radius vector (i.e., the segment connecting the Sun to the planet) sweeps out equal areas in equal times. As a result, a planet moves quickly when it is in the orbit closest to the Sun and slowly when it is far from the Sun. To deduce these laws, observations of formidable accuracy were necessary at the time. These had been conducted over the course of a lifetime, mainly by Kepler's mentor, the Danish astronomer Tycho Brahe. Printed in 1627, the report *Tabulae Rudolphinae* expounded on some of Brahe's most admirable observations about Mars (Fig. 1.3), which were then continued in Kepler's most important book, also one of the most important in the history of science of the sky: the *Astronomia Nova*.

In January 1610, Galileo Galilei aimed his telescope at the sky. That very night, a rudimentary tool in symbiosis with the eye of a great mind marked the birth of modern astronomy. Since then, the study of the stars has no longer been based on astrolabes and armillary spheres, but rather on optical telescopes, which have become increasingly diverse and powerful. Only in the twentieth century, with the extension of the field of astronomical studies to other bands of the electromagnetic spectrum and cosmic rays, and with the use of automatic probes, has there been a revolution in astronomy of comparable extent.

With the tools at his disposal, Galileo could not observe much of Mars; but it was only a matter of time before the optical and mechanical techniques would reach the necessary levels. Hence, the Dutch physicist Christian Huygens (1629–1695) first noted spots on the planet (shown in the figure at the start of this chapter), just enough to establish that Mars rotates completely in about 24 h, a fact confirmed shortly after by Gian Domenico Cassini (1625–1712). With Wilhelm Herschel (1738–1822), telescopes began to resemble the modern versions, at least in terms of size. As a result, astronomy, and not just the planetary field, began making extraordinary progress. Herschel, of German descent, after becoming Sir William in England, observed the occultation of a star from Mars. The phenomenon of occultation occurs when a star disappears behind the disk of a planet or the Moon. An occultation produced by the Moon is sudden and the star behind is seen to disappear immediately. However, the phenomenon occurs gradually with Mars. This is explained by the presence of a Martian atmosphere. Herschel also observed changes in the extent of ice sheets, interpreting them correctly as a seasonal effect.

The studies continued in the second half of the nineteenth century, but it is hard to find something new, even with continuously improving tools. We know today that a much more powerful telescope would be needed to detect the details of Mars, much finer than those available at that time. And even Hubble, NASA's space telescope, can spot only major features on the planet, such as the gorges of Valles Marineris. Throughout the history of science, when data start to run low, scientists are tempted to theorize, seeking new interpretations; and if the shortage of details becomes severe, sometimes data are just imagined (or sadly, in rare cases, even fabricated!) And so, a little fantasy begins to make its way into the maps of Mars. Local variations in darkness (called 'albedo' in astronomical and climatological terminology) could have a hundred different interpretations. But it is irresistible to make Mars into a brother planet of the Earth. Thus, Phillips identifies the black and white areas as lands separated by vast oceans (Fig. 1.4). A few years later, the British astronomer Richard Proctor draws up a new map, a synthesis of the knowledge about Mars. It shows, with great detail, seas, lakes, and rivers that separate boundless lands (Fig. 1.5). We know today that it would have been impossible to spot these details with the equipment available at that time. Perhaps the frustration of not being able to better observe the surface of Mars made some astronomers victims of biased observations and amazing performances. This is the way that one of the most famous blunders in the history of science made its way onto the record: the case of the Martian canals.

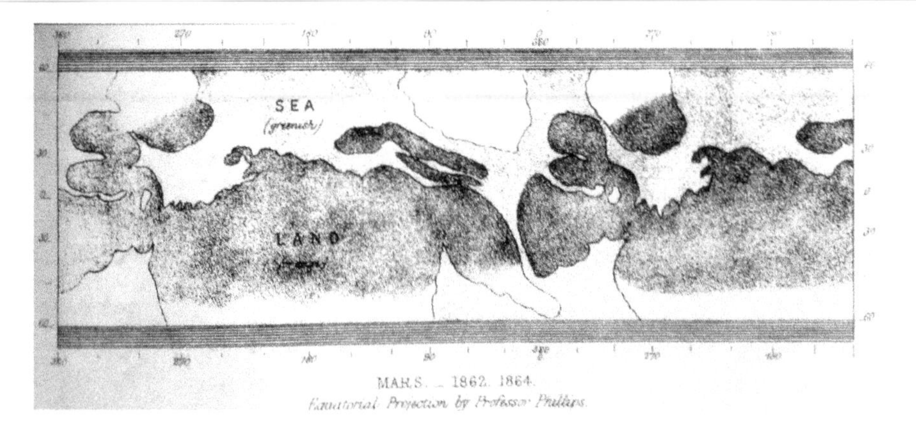

Fig. 1.4 The drawing by Phillips (1862–1864) published on the Proceedings of the Royal Society of London, 14, 42–46 (1865). Digital museum of planetary mapping, courtesy H. Hargittai, public domain

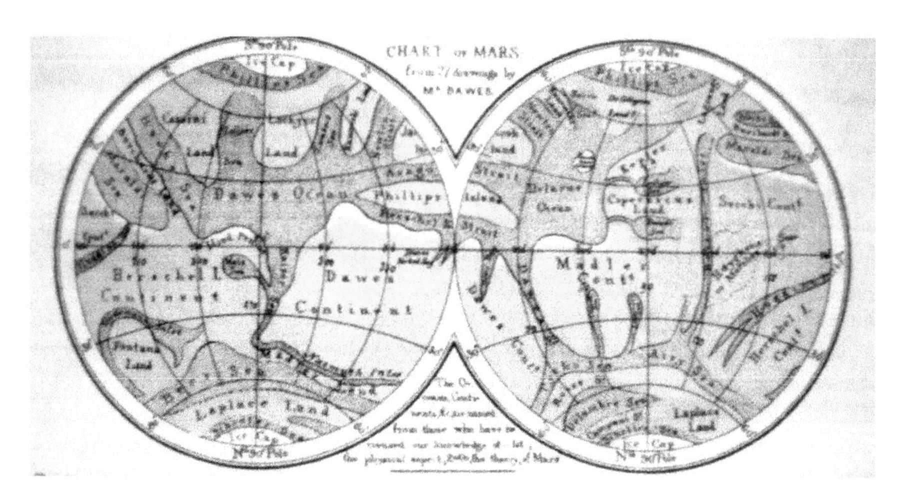

Fig. 1.5 The map by Proctor and Dawes of 1865 was regarded as the best in the pre-Schiaparelli era. Digital museum of planetary mapping, courtesy H. Hargittai, public domain

1.2 Schiaparelli, Lowell, and the Martians

The Golden Age of Telescopic Mars

It is ironic that Giovanni Virginio Schiaparelli, a scholar of many phenomena and a pioneer in the physical study of weather, an expert in stars and telescopes, and a man of great intelligence and modesty, also has his name linked to a prolonged observational error, partly due to his astigmatic sight. Born in the province of Cuneo (in the Piedmont region, northwestern Italy) in 1835, the most important Mars expert of the

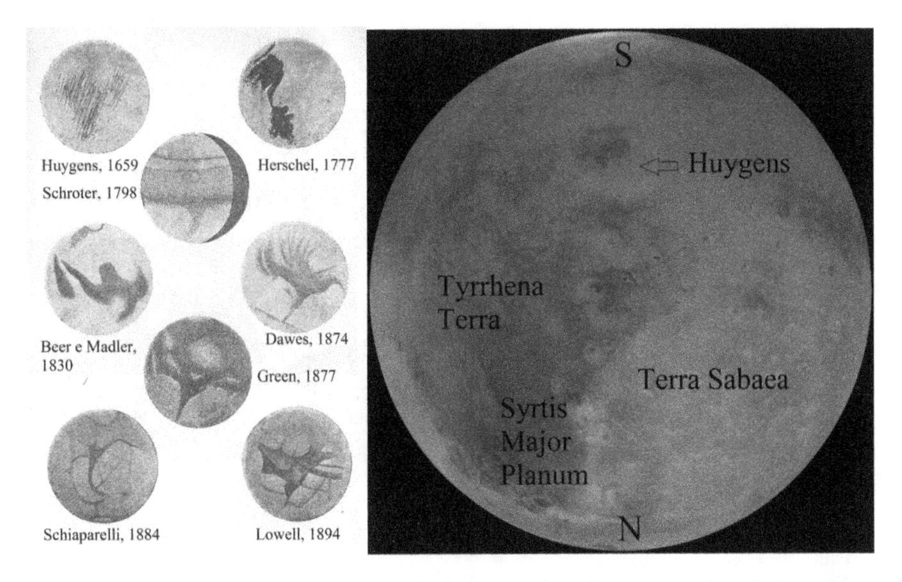

Fig. 1.6 Left: Syrtis Major (now Syrtis Major Planum) through two centuries of observations. Right: a modern image of Mars that includes Syrtis Major Planum in a Viking mosaic. For better comparison with telescopic observations, south is at the top. Second image courtesy of USGS Astrogeology, modified. Left: P. Hedegaard, Populaer Astronomi (public domain), mf; right: USGS and NASA, mf

pre-space period decided to become an astronomer when just 6 years old, struck by the charm of a swarm of shooting stars.

Before the introduction of photographic plates into astronomy, the sky was observed with the eye directly through the eyepiece of a telescope.[1] After leaving a career as an architect and engineer, the Italian astronomer made his observations from Milan with a 32-inch refractor Mertz telescope, a jewel for its time. The year was 1877, and Schiaparelli was already renowned for his important studies on comets. Like many other astronomers, he was eagerly awaiting the astronomical event of the year: the most favorable opposition of the Red Planet since the invention of the modern telescope.

Being of a greater distance from the Sun than the Earth, the planets Mars, Jupiter, Saturn, etc., are called the outer planets. An outer planet is in opposition when, during its orbital motion, it comes to be on the opposite side relative to the Sun. This is a configuration that occurs for Mars every 780 days (as the Mayans well knew) and during which observation becomes much easier, because the planet is located as close as possible to the Earth. But not all oppositions are alike: the Mars orbit is greatly elliptical, and so there are considerable variations in the distance between Earth and Mars oppositions. The opposition of 1877 turned out to be a success and

[1]Even the golden age of photographic plates is gone: it has long been replaced by digital systems such as CCDs, whose images can be reworked and enhanced on a computer.

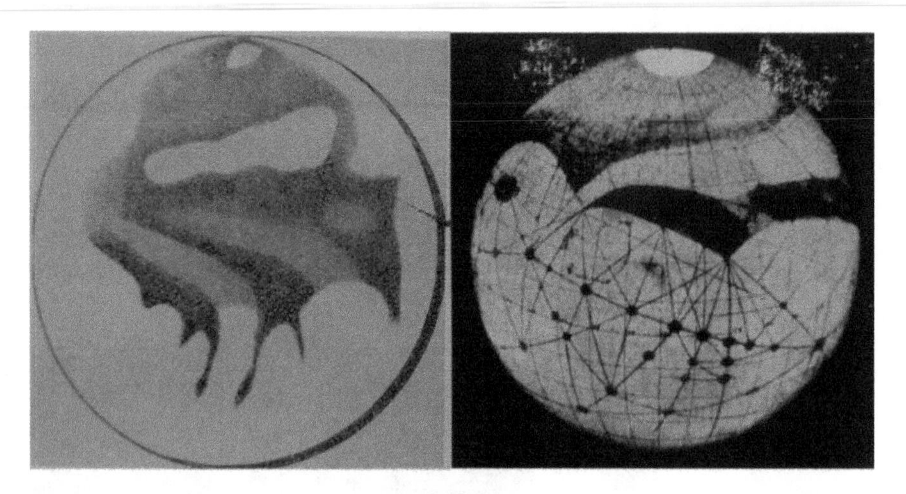

Fig. 1.7 Left: Mars as observed by Schiaparelli in Milan during the opposition of 1877. Channels are absent; they make their appearance in subsequent observations (right). Source unknown, public domain

Schiaparelli did not disappoint his colleagues: he promoted a map of Mars with new names, meant to describe features never seen before. To a strange white spot, he gave the name Nix Olympica: 'snow on Olympus.' We now know this to be the highest volcano in the solar system, renamed Olympus Mons. Noting that the polar ice caps are inclined with respect to the ecliptic at an angle similar to that of the Earth, he correctly deduced an inclination of the axis of rotation of the planet, and then the existence of seasons during the Martian year, just like on our planet. He also made important observations on the atmosphere of Mars. A few years later, he wrote, in his book *Life on Mars* of 1893:

> So much has been said about the polar snows of Mars, and how it is so incontestable that this planet, like Earth, is surrounded by an atmosphere that brings vapors from one place to another. Those are, in fact, snow precipitation of condensed vapors from the cold, which are then brought there; how are they brought, if not by way of atmospheric movement?

Today, we know that the "vapors" are transported carbon dioxide; but the observation of Mars' atmospheric instability is correct. Schiaparelli also describes something really peculiar: mysterious channels appear to cut the entire surface of Mars, joining areas of low albedo (Figs. 1.7, 1.8). The channels are also confirmed in the opposition 2 years later. What are they? Schiaparelli thought that they might be water channels. If water is stable on the surface of Mars, the pressure and temperature must be similar to those on Earth, otherwise the water would evaporate or freeze. Schiaparelli was convinced enough of the reality of water channels to deduce the atmospheric conditions of Mars from the "safe" presence of water, and not vice versa. He stated:

> The existence of an atmosphere charged with vapors was also confirmed by spectral observations, mainly those of Vogel, according to whom such an atmosphere would bear

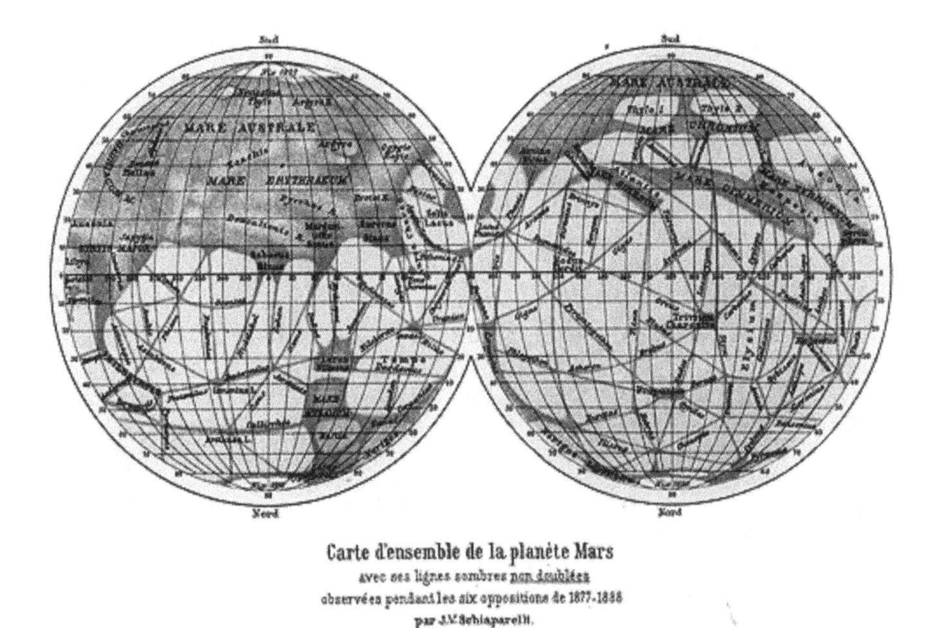

Fig. 1.8 Schiaparelli's "canali" are confirmed in this later map (1888). In most telescopic images, north is at the bottom, as the telescope reverses the image. Digital museum of planetary mapping, courtesy H. Hargittai, public domain

little difference from ours, and would above all be very rich in water vapor. This fact is supremely important, because it gives us the right to say, with much likelihood, that the seas of Mars and its polar snows are made up of water and not of other liquids. When this conclusion is confirmed beyond any doubt, another, no less serious one will also emerge: that the temperatures [on Mars], despite the greater distance from the Sun, are on the same order as those on Earth. Because if this is true, that which was supposed by some investigators, that the temperature of Mars is, on average, very low (50–60° below zero!), then it could no longer be assumed that water vapor is one of the main elements of the atmosphere of Mars, nor could water be one of the most important factors in its physical events; rather, that place should be left to carbonic acid or another liquid, the freezing point of which is much lower.

Schiaparelli, however, had no clue as to how the channels might form. In the absence of any evidence of alien technology, he initially favored a natural explanation. But in English, the word "canali" used in his Italian publications was translated as "canals," though "channels" would have been more correct. The semantic difference is significant, because while in Italian, a "canale" can be natural or artificial, and the word is suited to the prudence of Schiaparelli, in English, "canals" indicates an artificial structure. It is not the first time that a conception has been born from a translation error.

Many experts have been led to believe that there may be life on Mars. It was a wealthy American, struck by the recent discoveries of Schiaparelli, who passionately followed up the conundrum that he elicited. His name was Percival Lowell; when he decided to leave his diplomatic career to devote himself to Mars, he was still ambassador to Japan. He built an observatory in Flagstaff, Arizona, one of the observation spots in the whole world, owing to a lack of nearby villages, the dry desert climate, and an altitude of 2500 m above sea level. Lowell was convinced of the existence of the channels, and for many years, he kept on annotating countless details about them. The most astonishing phenomenon was channel splitting: at certain points, some of the channels appeared to have double, when only a few months earlier, the observations showed only one unit. How to explain such observations?

Lowell wrote thusly about the double canals:

> In good air the phenomenon is quite unmistakable. The two lines are as distinct and as distinctly parallel as possible. No draughtsman could draw them better. They are thoroughly Martian in their mathematical precision. At the very first glance, they convey, like all the other details of the canal system, the appearance of artificiality. It may be well to state this here definitely, for the benefit of such as, without having seen the canals, indulge in criticism about them. No one who has seen the canals well – and the well is all important for bringing out the characteristics that give the stamp of artificiality, the straightness and fineness of the lines-would ever have any doubt as to their seeming artificial, however he might choose to blind himself to the consequences.

So writes the more cautious, hesitant Schiaparelli, being caught between the fascinating idea of an alien civilization and the necessity of prudence as a scientist:

> The experience showed that it is not difficult to spot on the Moon, with the aid of the largest telescopes, a roundish object of half a kilometer in diameter, or a strip of 200 m wide. On Mars you can distinguish an object as round that is 60–70 km in diameter, and as a thin line a strip 30 km wide. The course of a river like the Po[2] would easily stand out on the Moon for almost its entire length, but none of the major rivers of the Earth would visible from Mars. And while on the Moon, a city like Milan (or even just Pavia) would be a subject well visible to us, on Mars, we could not hope to see even Paris or London, and it would be only vaguely possible to distinguish sizable rounded islands such as Majorca, or elongated islands the size of Candia and Cyprus.

In other words: to be visible from Earth, the channels should not only stretch northwise for thousands of kilometers and require an almost impossible technological effort; they should also be many tens of kilometers wide. But was there no shortage of water on Mars? Strange that these Martian engineers would build channels tens of kilometers wide to convey tiny gutters. Lowell was well aware of such difficulty, and so wrote in his book *Mars* (1895):

> The lines appear either absolutely straight from one end to the other, or curved in an equally uniform manner. [. . .]. The lines are as fine as they are straight. As a rule, they are of scarcely any perceptible breadth, seeming on the average to be less than a Martian degree, or about thirty miles wide. They differ slightly among themselves, some being a little broader than this; some a trifle finer, possibly not above fifteen miles across. Their length, not their breadth, renders them visible; for though at such a distance we could not distinguish a dot less than thirty miles in diameter, we could see a line of much less breadth, because of its length. Speaking generally, however, the lines are all of comparable width.

[2]The longest river in Italy.

Fig. 1.9 Mars according to Camille Flammarion in a Russian edition. Digital museum of planetary mapping, courtesy H. Hargittai, public domain

Thus, according to Lowell, length, and not breadth, is the reason why channels can be observed. As of 1895, Lowell was still dubious as to the nature of the channels, even though he clearly though they were related to life in some way. However, in the following years, Lowell became progressively dissatisfied with the mere observation and tentative hypotheses of Schiaparelli, and made no mystery of his unusual ideas. He plainly thought that there was a technological civilization on Mars, afflicted by a dry climate. Convinced that the channels were artificial structures of an advanced society, he initiated a saga of aliens that has nurtured science fiction for over a 100 years. That is what the channels are for: to transfer water from the poles to the equatorial regions, where it is absent. Every year, Lowell became more convinced of the reality of a Martian civilization, and the lack of competitors (his observatory was one of the best in the world) made him all the more enthusiastic. Looking at new twinning channels, where he had earlier observed just one (a splitting), he did not hesitate to support the idea, in an interview with the New York Times, that the Martian engineers must have been able to build a new channel within a few years.

Lowell was not alone in his speculation. The well-known French astronomer Camille Flammarion was also convinced of the existence of life in the solar system, and saw in Lowell's work a chance to materialize his belief in alien civilizations (Fig. 1.9). And so, the number of devotees of the idea of a Martian civilization grew, and it became a fashionable topic in high-class gatherings in the late 1880s.

Yet, some astronomers could not see the channels; perhaps, it was claimed, those observers were not skilled enough, or they were making use of weak instruments. This could not have been the case with Vincenzo Cerulli, however: he owned an observatory on a mound renamed Collurania near Teramo, now home to a section of

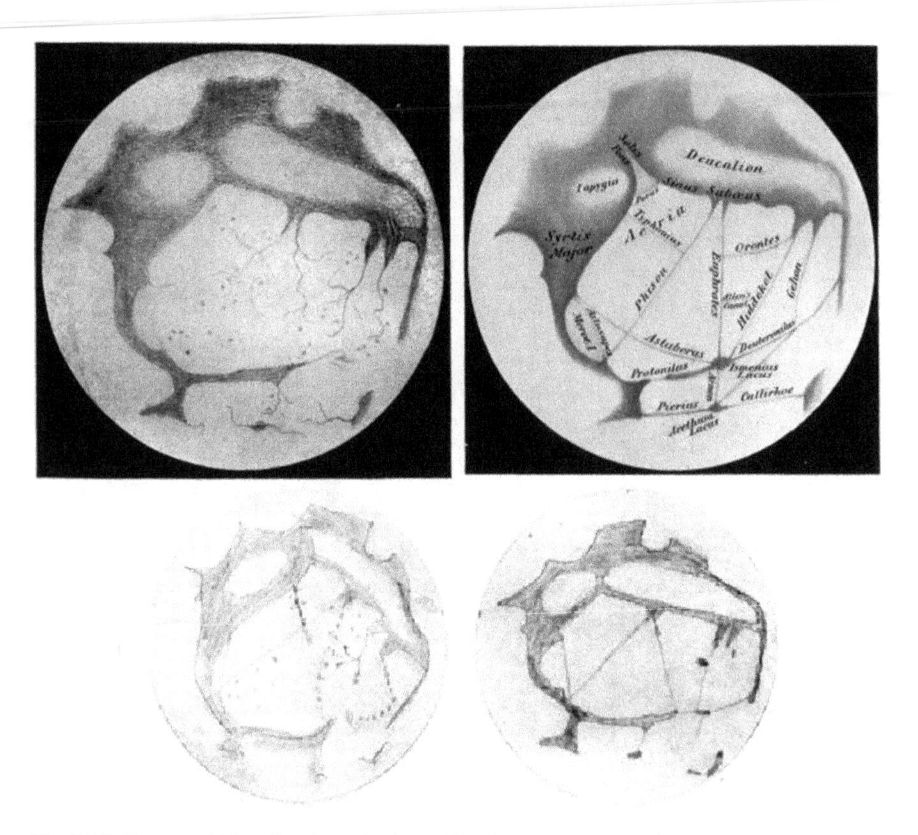

Fig. 1.10 Evans and Maunder showed a figure like the one at the top left (in which the "channels" are absent) to some schoolchildren. Looking at the figure from a certain distance, some of the pupils drew figures such as those shown at the bottom, where channels now join the upper to the lower black areas. At the top right, Mars is shown with channels such as Schiaparelli would have drawn. The similarity between the figure at the top right (Schiaparelli) and the ones at the bottom (unwitting pupils) is remarkable. It must be said, however, that Evans and Maunder led the schoolchildren in their perception with some snaking features. From: J. E. Evans and E. W. Maunder: Experiments as to the Actuality of the "Canals" Observed on Mars. M. Not. Roy. Astr. Soc. Lond. 63, 488–499 (1903). Courtesy Oxford University Press

the Italian National Institute of Astrophysics, and had excellent instruments at his disposal. Yet, he could not see them at all. This was at the end of the nineteenth century; roughly a decade later, in 1909, Antoniadi could not see any channels either, even with higher power instruments.

In June 1903, the British astronomers Evans and Maunder challenged the very existence of the canals, using decidedly unsophisticated instrumentation, namely a group of unsuspecting students. The two men provided a picture of the planet Mars as shown at the top left of Fig. 1.10. This was meant to represent the "true" Mars, i.e., Mars devoid of channels, as it should be given the hypothesis that the channels are an optical illusion. When the drawing was kept several meters away while the schoolchildren reproduced it, the children joined the stains with lines, just as Schiaparelli and Lowell did. And some even drew the twinning! This experiment on the perception of patterns supports a very simple theory: the brain, within the limits of extremely tough observational conditions, tends to unite spots with lines, creating the appearance of canals.

Martian Engineers or Martian Lichens?

In the absence of widely accepted evidence for or against, the charm of the Martians remained potent until the end of the nineteenth century (Fig. 1.11). Not everyone had been converted to the idea of the channels being an optical phenomenon. Certainly not Lowell, who died in 1916 still believing in the Martians. However, the opposition of 1909 elicited a change of belief in Lowellian supporters. Antoniadi, working out of the important Meudon observatory in Paris, began to draw Mars in a more natural way as a collection of spots without straight lines. This was nearly a return to the pre-Schiaparelli era, this time with improved telescopes. In the years around 1920, science remained open to the idea of a Martian civilization. In August 1924, during an opposition of Mars, radio stations were closed temporarily by the American Government, in the expectation that faint radio messages from Mars could be detected in the absence of background terrestrial noise.

Some scholars, while considering the channels and Lowellian speculation as mere fantasy, did believe in the existence of life on Mars. More than 20 years after the death of Lowell, many people still believed in the Martian channels, as well as in the vegetation associated with significant water masses (Fig. 1.12). In 1939, Pierre Humbert wrote, in a widely popular book entitled *From Mercury to Pluto*:

> The canal system stretches over the whole Martian globe; in the opaque regions, which we believe to be grasslands, one can see, as well as on continents, darker and narrower lines.

However, the wind was slowly changing: to many researchers, the original Lowellian interpretation of lines as artificial canals and the implied existence of an advanced civilization began to appear inadequate. Vegetation took over as the astrobiological dream. New solutions were also proposed as an answer to the too-wide-channels riddle, almost an aftermath of Lowellian ideas. Humbert again:

> Are these lines that we call "canals" real canals as our terrestrial examples? This appears to be unlikely: indeed, we cannot even imagine canals as wide as 30 km. What we see and call canals is actually the vegetated zone that forms seasonally along conduits that transport beneficial water.

Even in the 1950s, a few researchers attributed the ocher color of Mars to the presence of lichens. Lower life (if we may so call it), but life after all! The maps produced by Slipher at the Lowell observatory still showed the channels (Fig. 1.13). The channels, it was claimed, could no longer be dismissed as an optical illusion, since those maps were not purely visual as they were in the time of Schiaparelli, but rather were obtained by assembling more than 100,000 photographs.

Here, our story takes the journey of the Mariner 4 in July 1965.

We return, then, to those twenty-two plates taken by Mariner 4.

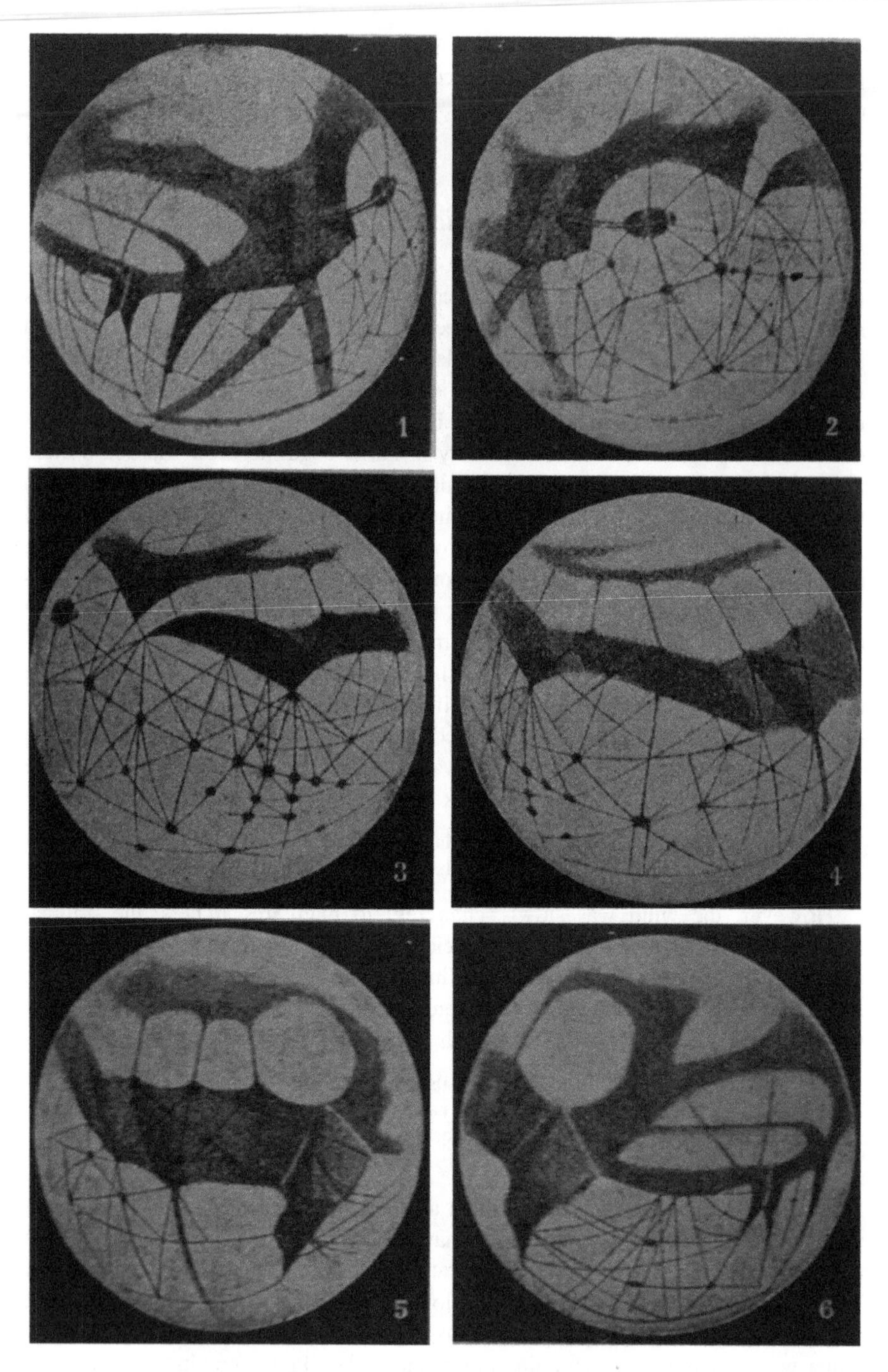

Fig. 1.11 A complete map of Mars as proposed at the end of nineteenth century. Each picture shows the appearance of the planet rotated by 60°. P. Hedegaard, Populaer Astronomi (public domain)

Fig. 1.12 In the 1930s, the view prevailed that the “canals” on Mars were not artifacts, but rather natural watercourses hosting a variety of vegetal life. Image 16232428 FOTOLIA/Archivist

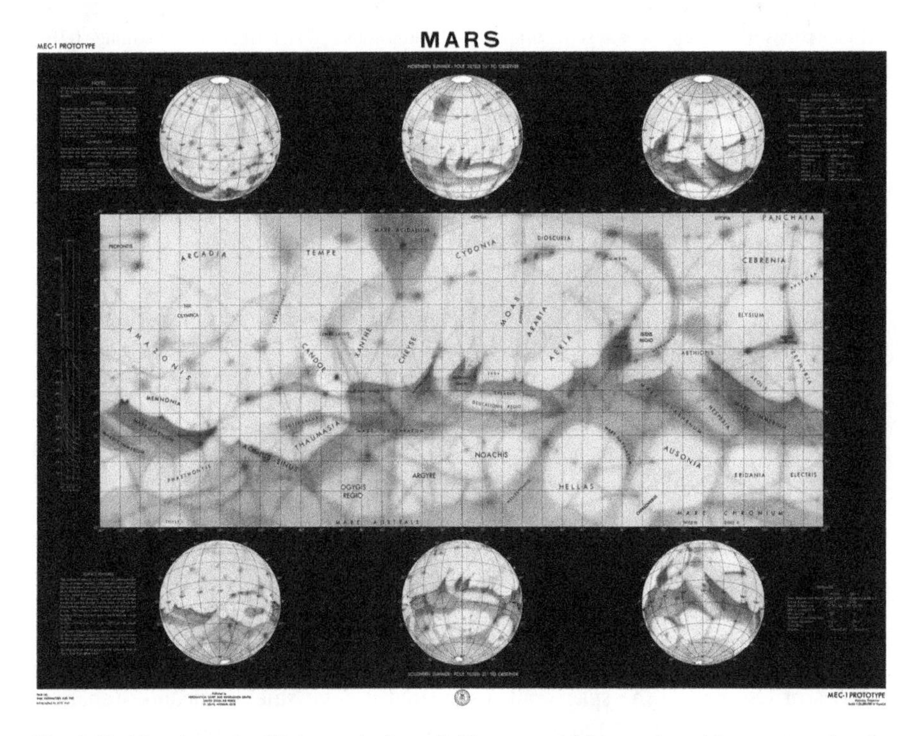

Fig. 1.13 Map drawn by Slipher at the Lowell Observatory. Maps such as this one, reporting the channels, were still being used in 1962 during the Mariner exploration. Courtesy of the Library of Congress, USA. Courtesy library of Congress, USA

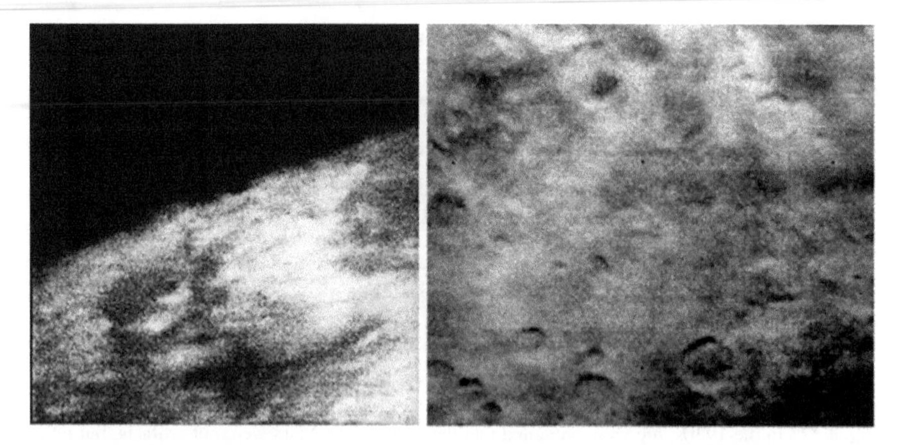

Fig. 1.14 The first close-up images of Mars taken by Mariner 4 in July 1965. The one on the left shows cloud systems between the Elysium and Arcadia Planitiae (image 01D); the photo on the right identifies the region southwest of Amazonis Planitia, centered at −13.5°, 174.2° (image 07B). The early images did not record any canals, confirming their psychological origin. Courtesy of NASA. Mariner 4, NASA

1.3 Modern Mars

Mars Seen by Mariners and Vikings

To compare the new photos with what was already known from ground-based observations, scientists had the maps drawn by Slipher (at the observatory founded by Lowell), which were not particularly different from those compiled by Schiaparelli nearly a century earlier.

But here is what Mariner saw (Fig. 1.14): craters on a barren landscape, with no evidence of water, let alone the presence of canals. Mars was more like the Moon than the idyllic planet imagined by Schiaparelli and Lowell! Thus, the excitement of the technical success of the mission was followed by scientific disappointment: Mars looked like a dead planet.

The surprises were not over. A host of new ideas about Mars was elicited by another spacecraft. Mariner 9 returned with more than fifty thousand images, of much improved quality compared to Mariner 4. Where Schiaparelli had noticed the "snows of Olympus," the spacecraft discovered a huge shield volcano. Renamed Olympus Mons, with its 21 km of height, it is the tallest volcano in the solar system. Hovering over the equatorial regions, Mariner 9 discovered a system of gorges 4000 km long and 8 km deep: baptized the Valles Marineris, they are among the few features that coincide with one of Schiaparelli's channels. We will discuss these incredible and enigmatic valleys in detail.

If the expectations of 60 years earlier had been completely disappointed by the absence of any Martian civilization, Mars was now once more at the center of studies, and it was there to stay. On August 20, 1975, a new spacecraft left Earth

Fig. 1.15 The launch of the Viking II from Cape Canaveral on August 20, 1977 by means of a Titan/Centaur rocket. Part of PIA1480 image. Courtesy of NASA/JPL. NASA

for the Red Planet (Fig. 1.15). It was called Viking 1 and had technical capabilities far superior to the Mariners. The mission not only provided a spacecraft designed to orbit around Mars, but also a scientific unit meant to touch the surface of the planet, called a lander. The journey of 11 months ran perfectly, until a decision had to be made as to the position at which the lander would be released. This was a delicate issue, because a surface that was too rough could topple the lander, throwing away years of work. Image analysis from the orbiter could have helped, but it would take time. Having missed the symbolic date of July 4, American Independence day, the staff at NASA opted for July 20. The chosen location was a spot on the Chrise Planitia, near the volcanic Tharsis area. The powerful thrusters needed to guarantee a soft landing were guided perfectly by radar: it was the first time that a human artifact had reached the surface of another planet.

The first photos showed a rocky surface, surrounded by a sea of tennis court-colored sand (Fig. 1.16). The Viking 2 sister mission, launched at Cape Canaveral 20 days after its predecessor, landed on Utopia Planitia, a region at higher latitude

Fig. 1.16 One of the first images from the Viking 1 lander. It shows sand dunes similar to the desert dunes on Earth, peppered with angular blocks. Image PIA00383. Viking 1 (NASA)

immersed in the desolate plains of Northern Mars. Four billion years ago, an asteroid of more than 200 km in diameter hit the region, levelling the area and creating one of the largest impact basins in the solar system.

The harvest of scientific information of the two Vikings, both in terms of photos from the orbiters and data from the landers, was enormous. The probes had studied details of the surface, such as giant water channels, impact craters, and landslides. They cataloged the Martian volcanoes, from the huge ones like Alba Patera (now Alba Mons) to the smaller ones. The composition of the atmosphere was precisely determined with a mass spectrometer (an instrument that measures atomic mass) and a gas chromatograph; many anomalies in isotopic ratios relative to Earth were found. These were the first concrete and precise results obtained from Mars. Often, the Vikings are remembered for another fascinating and controversial result: the possible detection of chemical products of alleged Martian life. Viking 1 apparently found compounds of carbon and nitrogen that may be a signature of life. But with much regret, the experiments with the lander turned out to be ambiguous; since those missions, the issue has been filed as a spurious result biased by the limits of the experimental apparatus. Thus, the question persists: is there life on Mars? This will be a major topic for years to come. The large volume published in 1992 by the University of Arizona stood as the first summary on the science of the planet Mars, still only in its infancy up to that point.

A Long Interval and the Return to Mars

The enormous success of the Vikings paradoxically created a sense of emptiness in the Martian strategies of NASA. Mars had been conquered. Undoubtedly, it was a very interesting planet, rich in geological landscapes such as volcanoes and outflow channels. But there seemed to be no life. As with the Moon, hectically explored within a few years by the Apollo manned missions and afterwards neglected, there also began a long-lasting period of little interest in the Red Planet. There was no lack of ideas, proposals and projects. However, the absence of a clearly defined and important goal, like the search for extraterrestrial life, slowed enthusiasm. Budgetary issues, exacerbated by competition from other large projects (like the Space Shuttle, which resulted in only partial success), forced the NASA missions staff into austerity conditions, and misfortune did the rest. Thus, the Mars Observer mission of 1990 failed to enter Mars' orbit and was lost.

Only in September 1996 did the Mars Global Surveyor (MGS), followed by the Pathfinder mission of July 4, 1997, succeed perfectly. These missions were also important from a technical viewpoint. On the plains of Utopia Planitia, the Pathfinder deployed the first rover on a planet other than the Earth, albeit a toy-sized one: it was called the Sojourner. To put the spacecraft in orbit around Mars, MGS used, for the first time, the technique of aerobraking (Chap. 2.1). Pathfinder also tested the aerobag, a cocoon of air chambers protecting the descent of the Sojourner rover, a technology borrowed from Russian missions.

Other countries have joined the Unites States and Europe in Mars exploration. Russia (formerly the Soviet Union), traditionally America's rival in the space rush, has, however, been very unlucky. As far as we know—many failures remained secret for a long time—the first mission, Mars 1, was launched in November 1962, but communications were lost with the probe while it was still far from the Red Planet. Successive missions also represented a list of failures.[3] Even Mars 3, despite an apparently successful landing, only succeeded in sending data for a short time period, mere seconds. Dozens of photographs and other data proved very valuable in revealing high mountains and craters; they also confirmed the atmospheric pressure as being at about 6 millibars (or 600 pascals). Initially, it was suspected that the problem with Mars 3 was a furious storm taking place on Mars at the moment of the parachute deployment, which proved fatal for a completely automatic mission. However, later studies identified a different culprit: the parachute itself. Mars 3 did land softly, only to be enveloped by its own parachute, which prevented any communication. After the failure of the Mars 4, 5, 6 and 7 and two spacecraft dedicated to the study of the moon of Mars, Phobos, the Russians decided to dedicate their efforts, thus far successfully, to the study of Venus.

Technical Box 1: Sending a Spacecraft to Mars

The first step needed to send a spacecraft is to transcend the Earth's gravity field (Fig. 1.17). Physics tells us that the energy needed per unit mass to escape the Earth is just the product of the Earth's radius times the gravity acceleration at the surface. This gives energy of about 62 MJ for each kilogram sent out into space (this is the energy released by about 2 l of gasoline).[4]

(continued)

[3]In 1971, it was Mars 2 and Mars 3's turn, each consisting of one orbiter and a lander. After reaching orbit, communication with Mars 2 was lost. Mars 3 experienced engine problems and became drawn into the wrong orbit, one way too far from Mars. At nearly 5 tons, the landers of Mars 2 and Mars 3 were heavy. When the time of landing came on November 27 and December 2, 1971, communication with Mars 2 was lost immediately after penetration into the atmosphere, and only the Mars 3 lander made it to the surface of the planet.

[4]MJ, or mega joule, is 10^6, or one million joules; 10^6 means a "1" followed by six zeroes, i.e., one million.

Technical Box 1 (continued)

After the spacecraft is outside of the influence of the Earth, one has to overcome the gravity of the Sun. The Earth is closer to the Sun than Mars is. Thus, we have to provide energy in order to reach an outer planet, and a theorem of mechanics tells us how much. It follows that, for each kilogram of a body in orbit around the Sun at the Earth's distance, a gravitational energy is associated on the magnitude of −445 MJ (the negative sign reminds us that the body is bound to the Sun). This is the gravitational energy of the Earth itself, for each kilogram of mass. For a body at a distance of that of Mars, this energy is higher, because Mars is farther from the Sun than the Earth is. The figure is: −292 MJ per kilogram (the absolute value of the number is smaller, but the energy is higher due to the negative sign).

Thus, in order to reach Mars, one has to provide the difference between these two energies, or 445−292 = 153 MJ per kilogram of spacecraft, to which we have to add the energy needed to escape the Earth that we had calculated earlier, 62 MJ per kilogram. The Mars Science Laboratory (MSL), which provided the Curiosity rover at Mars, weighed it at 521 tons. This gives a total theoretical energy on the order of 10^8 MJ. To put this number into context, this is the energy produced in 1 day by a relatively small nuclear power plant of 1000 MW.

What orbit should the spacecraft follow to reach Mars? This problem was studied by the German engineer Walter Hohmann in the 1920s, in advance of the first space flight by nearly 40 years. The best transfer orbit turns out to be an ellipse whose perihelion touches the Earth's orbit, and the aphelion is on Mars' orbit (Fig. 1.17). This is the orbit that is economically more convenient. On a typical mission, the rocket is fired to escape earth's gravity field. Then, in a second stage, the spacecraft is propelled parallel to the Earth's orbit to make it follow the elliptical Hohmann orbit (red in the figure). After about 8 months, Mars is reached on the Hohmann's orbit, but the energy of the spacecraft is too low to stay in the Martian orbit. In practice, the spacecraft is travelling at lower speed than Mars on its orbit. If no adjustment is made at that point, the spacecraft will continue following the orbit shown in red in the figure, thus abandoning Mars again. However, if a rendezvous mission is planned in which an orbiter will stay close to the planet (and perhaps send a lander or rover down to the planet's surface), a third rocket should be fired when the spacecraft is close to Mars, along the direction of Mars' velocity. This would give the probe the extra energy necessary to follow Mars' orbit, avoiding the return orbit. A smart technique is to let the spacecraft get so close to Mars that the planet's atmosphere will assist in capturing it, a procedure called aerobraking.

The escape velocity is the one required for a body to be thrown from the Earth's surface to reach space. Technically, this is obtained by equating the gravitational energy with the initial kinetic energy of the body, from which it

(continued)

Technical Box 1 (continued)
follows that the escape velocity from the Earth's surface is 11.2 km/s, while for Mars, it is 5.03 km/s. This definition will be useful in dealing with many aspects of Mars, from the velocity of asteroid impact to the fate of Mars' atmosphere. Sending a spacecraft to Mars requires a huge technical and economical effort, which has been celebrated with commemorative stamps (Fig. 1.18).

1.4 Mars in Its Orbit

Mars is the fourth planet from the Sun (Fig. 1.19). Its distance from the Sun is 50% greater than that of the Earth from the Sun (Fig. 1.20). Our star appears as a disk of 0.33 degrees (0.33°). From the Earth, the Sun is about half a degree across, and 2.25 times brighter. Two orbital characteristics of Mars are similar to those of our home planet. The Martian day (called sol) is only 37 min and 22 s longer than the 24 h of the terrestrial day. And the obliquity or the inclination of the axis of rotation with respect to the plane of orbit is also comparable: 25°19′ versus the terrestrial value of 23°26′. Thus, sitting on a Martian hill, we would perceive the passage of the day in a manner similar to that which we experience on Earth. In addition, Mars is also a highly seasonal planet, with long seasons of more than twice the duration of those on Earth.

With an eccentricity of 0.093, the elliptical orbit of Mars deviates from the circle by a much greater degree than that of our planet (the eccentricity of Earth's orbit is only 0.017, almost a perfect circle). This observable eccentricity allowed Kepler to discover the first two laws of planetary motion. This small but significant flattening

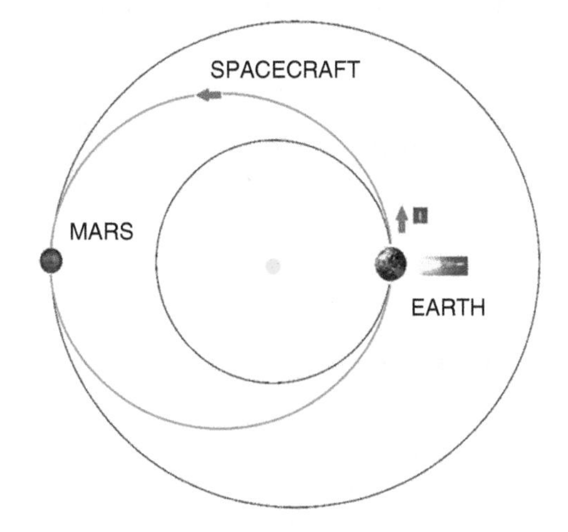

Fig. 1.17 Hohmann transfer orbit from the Earth to Mars. FVDB

Fig. 1.18 Various commemorative stamps issued by different nations celebrating the American missions Mariner 4 and the two Vikings (first four stamps) and the Russian missions Mars 2 and Mars 3 (last four). The parachutes of Mars 3 are drawn as correctly deployed; however, it is believed that they actually wrapped themselves around the spacecraft, preventing any communication with the Earth. FOTOLIA

Fig. 1.19 Left: Mars is the fourth planet from the Sun. Planets and orbits not to scale. Right: Mars is about one half the size of the Earth. Also, the Moon is shown to scale. Left: images 143598796 and 143598571 FOTOLIA/mkarco

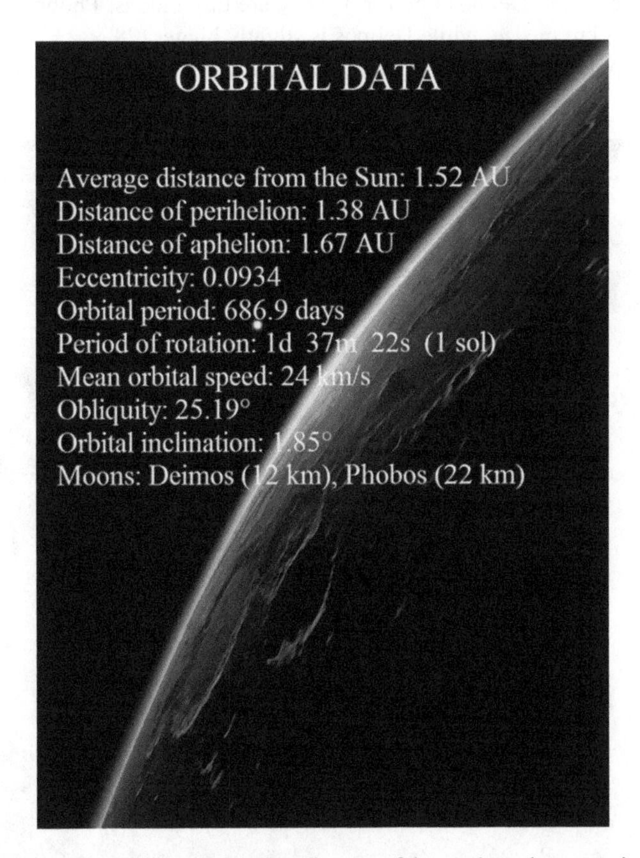

Fig. 1.20 Orbital characteristics of Mars. The diameter of the two irregular moons is calculated as the average between the major and minor axes. NASA, mf

of the orbit causes Mars to come closer to the Sun at perihelion with a minimum distance of 1.381 UA (a UA, the astronomic unit equal to 149.5 million kilometers, is the average distance between the Earth and the Sun), while the aphelion is far more distant: 1.666 UA. It seems a small difference, but at aphelion, Mars receives 45% less solar radiation than at perihelion. Since the Martian southern summer falls when the planet is almost at perihelion, an imbalance sets in between the radiation received by the North Pole and that received by the South Pole. The consequences on the climate of Mars are important, and will be discussed. As for the Earth, the eccentricity of Mars changes over time during a period of about one hundred thousand years. This has important consequences on the long-term climate of Mars, most of which has yet to be understood. The sidereal period of revolution is almost 700 days.

Because planets farther from the Sun travel more slowly, the average orbital speed of Mars is less than that of the Earth, or 24 km/h. The escape velocity of 5 km/s (Technical Box 1), higher than that on the Moon, has important implications for return trips to Mars, as it requires a large driving force. The two moons of Mars were discovered in 1877 (Fig. 1.21). Taking inspiration from the horses pulling the chariot of the god Ares, the astronomer Asaph Hall named them Phobos (from the Greek word for "fear") and Deimos ("terror"). They are tiny moons: Phobos is a boulder 27 × 22 × 18 km in size, while Deimos is slightly bigger (28 × 23 × 20 km). The irregular shape of these heavenly bodies is a consequence of their small size, which

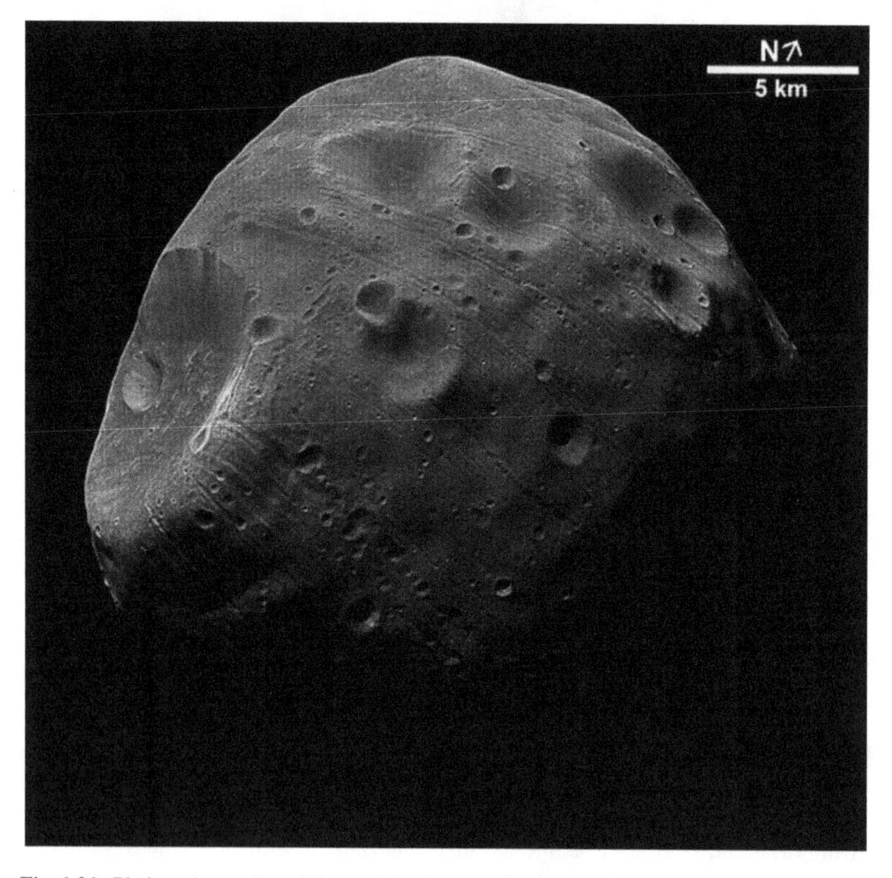

Fig. 1.21 Phobos, the smaller of the two Martian moons, is gradually getting closer to the surface of the planet. In a few million years, the tidal forces will exceed the strength of the rock, and the moon will disintegrate. Image ESA/DLR/FU Berlin (G. Neukum). ESA/DLR/FU Berlin (G. Neukum)

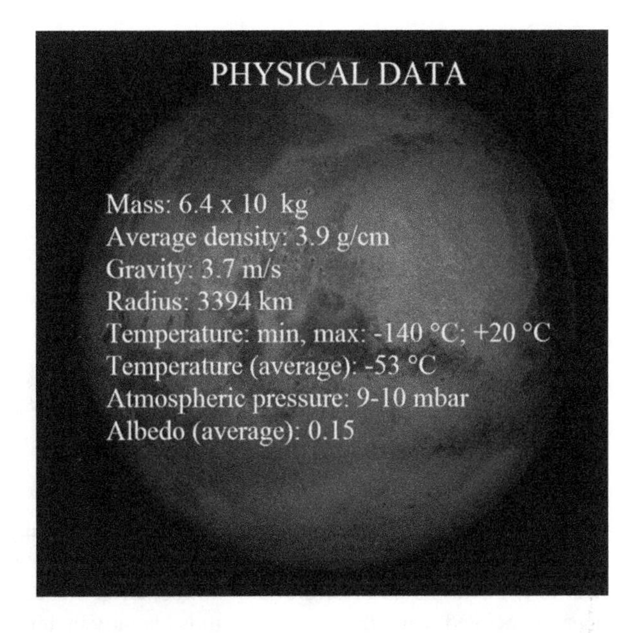

Fig. 1.22 Physical characteristics of Mars. NASA, mf

induces a very small gravity field on the surface, about 1/100,000 that on Earth. Only larger planetary bodies create a gravity field strong enough to force their surface to assume a spherical contour.

Being very close to Mars (only 6000 km, or 50 times closer than the Moon is to Earth), Phobos circles the planet very quickly. As a result, it orbits faster than the rotation of Mars on its axis. A unique case in the solar system, Phobos rises in the west and sets in the east. It is getting closer and closer to the Martian surface, and will eventually disintegrate in a few million years, perhaps creating an impact crater 200 km in diameter.

Also, many of the physical characteristics of Mars have been known for several decades (Fig. 1.22). Mars has a mass equal to only one tenth that of Earth, a result of the smaller average radius (3394 km, in comparison to the 6378 km of the Earth, Fig. 1.19) and also of the lower average density (3.9 g/cm^3 compared to 5.5 g/cm^3 for the Earth). Consequently, the Martian gravity acceleration, which is proportional to the mass of the planet and inversely proportional to the square of the radius, is slightly greater than a third of its terrestrial equivalent: its magnitude is 3.70 m/s^2 on average. A person would, of course, well perceive the lower gravity, even while walking normally. Apparently, lower gravity may appear to be a positive characteristic, as it promotes freedom of body movement. The bar for the high jump championships on Mars should be set to heights between 5 m and 6 m, and the volleyball net would need to be as high as 3.5 m, to be compared with the official 2.43 m on Earth. However, in the long run, the reduced gravity would affect muscular efficiency to the point of causing them to atrophy; it is truly a serious concern for human exploration of the planet.

Because Mars receives less light from the Sun, and since the greenhouse effect is negligible, temperatures on Mars are much lower than on Earth. Average temperatures drop to 53 below zero; only at the equator could one occasionally enjoy +20 degrees of temperature. The very thin atmosphere implies ground pressure much lower than that of our planet: 9–10 millibars versus 900 millibars. Among the many implications of this, a particularly notable one is the fact that water is not thermodynamically stable, and therefore, despite the low temperatures, we would see the water in a glass boiling and then freezing.

1.5 The Rocks on the Martian Surface

The Reddish Color

The reddish color of the Martian surface has linked the planet to astrological and mythological interpretations throughout its history. Because this color evokes the heat of fire and the color of blood in almost all civilizations, Mars has consistently been associated with the bloodiest gods, a violent temper, and highly dramatic events. Who knows how many political and military decisions were made based on the position of the Red Planet on the ecliptic? However, the reason for this red—actually stained rust—color is quite interesting, dipping, as it does, into the peculiar evolution of Mars. A newspaper from 1952 enlightens us:

> Mars is red, a color that comes from lichens that cover the surface. It's a flat red and cold world, cut by the broad outlines of its melancholy channels.

Today, we know that the red color is due to the presence of an iron oxide surface. Because the volcanic rocks that cover the surface of the planet are iron-rich, this element abounds on Mars. Iron is common in basaltic rocks, the most abundant volcanic rocks on Earth. It is now believed that the surface of Mars was originally, indeed, since its formation, rich in basaltic magma that has been there for billions of years. On Earth, as well as on Mars, basaltic iron derives either from a mineral of the pyroxene group, such as augite, or from olivine, a beautiful green mineral that is also used as a gem. Basalts are altered by weathering, a process that occurs only in the presence of oxygen. This process combined the iron originally contained in the augite and olivine minerals with oxygen to form hematite, an iron ore. Dark in crystalline form, hematite becomes reddish when reduced to a fine powder, hence the name referring to blood (Figs. 1.23, 1.24). It is now believed that the red color of Mars is due to a layer of hematite in finely powdered form, perhaps ground to nanoparticle size. Not a deep layer, but just a surface of a few decimeters, like sugar topping on a cake.

On Mars, the oxygen is not chemically available, as it is combined in rocks and partly bound with carbon to form an atmosphere of carbon dioxide. What process, then, helped forming the hematite on Mars, if the environment is so oxygen-poor? Related to this problem, there are several other questions. Why is Mars composed

Fig. 1.23 The red color of Mars is the result of iron oxides on the surface. Image STScl-1999–27, Hubble space telescope. (NASA)

Fig. 1.24 Hematite appears black when in crystalline form (left). However, notice the red-brown spots on this specimen, where the mineral is powdery rather than crystalline. The reddish-brown color becomes even more evident on a stripe on white marble (right). Today, the Martian oxygen of the atmosphere is bound to carbon, making up the carbon dioxide molecule, but it must have once been available in an aqueous environment. FVDB

almost exclusively of basalt on the surface, while the Earth is much more differentiated in its many different rocks? And why is there so much iron on the surface of Mars, while the terrestrial iron is concentrated in the interior of our planet? Scientists are trying to address these and other questions on the origin of the Martian rocks. We can anticipate that some processes of hematite formation may take place in the absence of free oxygen, provided that water is present, too, and this is yet another clue to the presence of water on ancient Mars.

The Final Answer

Nowadays, we can finally answer the chief question that all of the pre-Mariner observers must have asked themselves: is there a correspondence between the telescope image and the morphology of the planet? Do mountains, volcanoes, and northern plains reported on a modern map correspond to Schiaparelli's spots, Lowell's channels, or the *Maria* of Slipher? Fig. 1.25 shows the last map drawn by Slipher in the top panel (the same map has already been shown in Fig. 1.11) compared to a modern MOLA map showing the elevations with a shade effect (figure in the middle; this is essentially the black and white version of the frontispiece figure). There is indeed some correspondence, which is, however, limited to a few giant morphologies: the impact basins of Hellas and Argyre Planitiae to the south, the Olympus Mons volcano, formerly called Nix Olimpica (northwest), Valles Marineris (equatorial Mars), and a few others. The dark areas on Slipher's map appear to have no correspondence with modern elevation maps. However, if we overlay the local albedo as measured by modern instrumentation on the MOLA map (bottom figure), we finally understand what pre-Mariner observers actually saw: superimposed onto the giant morphologies, the contrast between the light and dark areas of the planet becomes the dominant theme. Comparing the top and bottom figures of Fig. 1.25, the correspondence between the dark and light areas becomes apparent. In short, observers saw variations in the albedo superimposed onto a limited number of giant geological structures.

We should not belittle the achievements of the scientists working before the space age. Their measurements of some of Mars' properties were actually quite accurate. However, as always occurs in the history of science, there is a desire to give a rational meaning to peculiar facts that otherwise would be hard to explain. Initially, some astronomers interpreted the dark areas on Mars as oceans and the white ones as continents, and thus maps with continents engulfed by vast oceans flourished. Schiaparelli contested these ideas, on the ground that the atmosphere was too tenuous to ensure the thermodynamic stability of water. Ironically, by performing incorrect observations of the putative channels, he ultimately re-introduced the idea that water was present on the Martian surface. At any rate, no matter what these darker spots were, they provided a very visible marker on the surface of the planet, and thus were instrumental in measuring the orbital characteristics collected in Fig. 1.12. Some of these were known with good approximation one century before space exploration.

Addenda

- The Latin name Mars is of unknown origin; it is probably not Indo-European.
- The eccentricity of the planets has been significant for the entire history of astronomy. Examination of the eccentric orbits of Mars and Mercury were of great importance, respectively, for the development of the heliocentric system and general relativity.

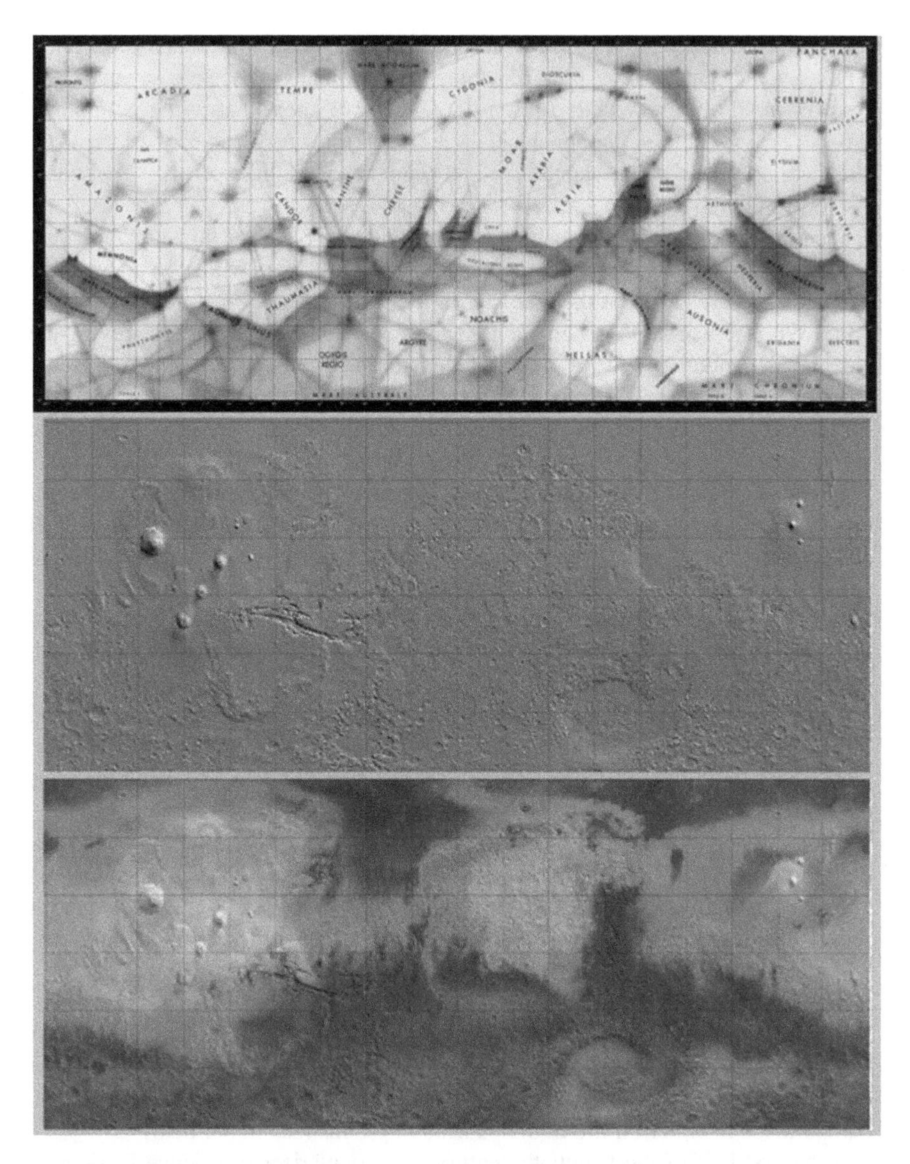

Fig. 1.25 Comparison between the map by Slipher (top), the altitude MOLA shaded map (middle) and the MOLA shaded map with superimposed albedo (bottom). Note the correspondence between the first and last images. Thus, the dark areas, which were correctly reported by many early observers, do not necessarily correspond to geological features, but rather follow the superficial distribution of powder-sized particles. Top: courtesy library of Congress, USA, middle and bottom: MOLA images on JMARS platform

- One of the features most visible to telescopes is the region known as Syrtis Major. For several centuries, it has been the testing ground for serious amateur and professional observers alike (Fig. 1.6). Today, we know that Syrtis is home to very dark volcanoes, but in general, the spots on the surface of Mars show deviations in albedo (the capability to reflect or absorb light) more than geological structures.
- The next very favorable opposition of Mars will take place on July 27, 2018. At a distance of 0.386 astronomical units from our planet, Mars will subtend an angle of 24.1 s of arc (1.3% of the diameter of the Moon).
- In 1898, the astronomer Samuel Laland speculated that the telescopes of the twentieth century would reveal the towns on Mars in detail, as well as ships and ports of the Martian civilization.
- The English composer Gustav Holst wrote one of the best pieces of music dedicated to Mars at the opening of the vast symphonic poem "The Planets" (1914). A pounding, fearful rhythm is attained by placing a triplet at the beginning of each bar throughout most of the piece, except for the middle section. Had Holst simply appropriated the theme of "Mars, bringer of war" for the sake of mythology, or did he foresee the disaster of World War II (see figure below)?
- Before the Mariners, the surface of Mars was known with a resolution of less than 100 km. Mariner 9 improved the resolution to 1 km. Modern images taken with the latest probes can show details of less than a meter.
- Because of the high eccentricity of the Martian orbit, the northern spring, summer, autumn, and winter last, respectively, 199, 181, 145, and 160 days. On Earth, the difference between seasonal duration (91 days for each season) is minimal.
- The existence of two moons orbiting Mars, their small diameter, and even their distance from the planet was fortuitously predicted with considerable accuracy in the novel "Gulliver's Travels" by Jonathan Swift, published in 1726. The moons were discovered in 1877 by Asaph Hall, an American joiner turned astronomer, more than one and a half centuries after the novel (see figure below).
- Despite his blunder with the canals, Lowell has remained an important astronomer: he built an outstanding observatory (where, among other things, the redshift of galaxies was discovered) and indicated where to look for "Planet X," where Pluto was found. He was also one of the earliest scientists to understand the importance of water on life in the planetary systems (see figure below).
- In the first half of the twentieth century, the popularizer of science Desiderius Papp speculated as to how a hypothetical humanity would evolve on Mars. The reduced gravity would raise Martian humanoids up to five meters, with a skeleton less robust than ours, fur, and very hot blood. Reduced verbal communication due to the low air density would promote light signals as a substitute.

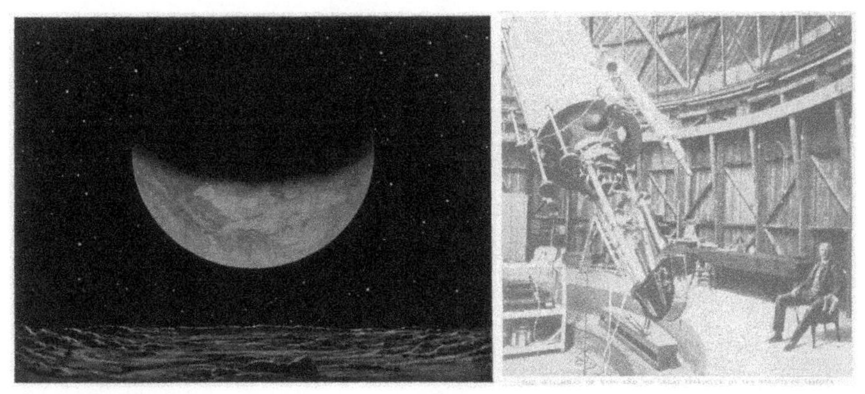

The symphonic poem “The Planets” by Gustav Holst starts with “Mars, the bringer of war”; an artist’s conception of Mars as seen from Phobos in a drawing from the 1930s; Percival Lowell sitting inside the Lowell Observatory. Public domain, FOTOLIA image 162407887/ Archivist, image 162423501 FOTOLIA/Archivist

Chapter 2
History and Physiography of Mars

Observing Mars from a spacecraft in polar orbit, the plains of the northern hemisphere would appear smooth and with few features except for the polar ice cap. As the spacecraft shifts to the south and the planet spins in front of our eyes, huge volcanoes would appear, well recognizable even at distances of thousands of kilometers. Approximately at the equator, we would observe the long canyons of Valles Marineris. Only continuing to the southern part of Mars would we notice a mountainous surface completely crammed with craters, reminding us a little of our Moon. This division of the planet into two morphologically different hemispheres has been defined, by some, as one of the most amazing discoveries in planetology. It still appears extremely enigmatic today. Before looking at the morphology of Mars in more detail, it is useful to know its main features. In the pages that follow, we shall start with the strange dichotomy of Mars, see the most extensive landscapes of the Red Planet, and then go into its history. We will assess the current knowledge on the planet's interior, the mystery of the disappearance of the magnetic field, to which the destiny of life on the planet could be tied, and shed light on its main geographic features and history as deduced from the images.

F. V. De Blasio, *Mysteries of Mars*, Springer Praxis Books,
https://doi.org/10.1007/978-3-319-74784-2_2

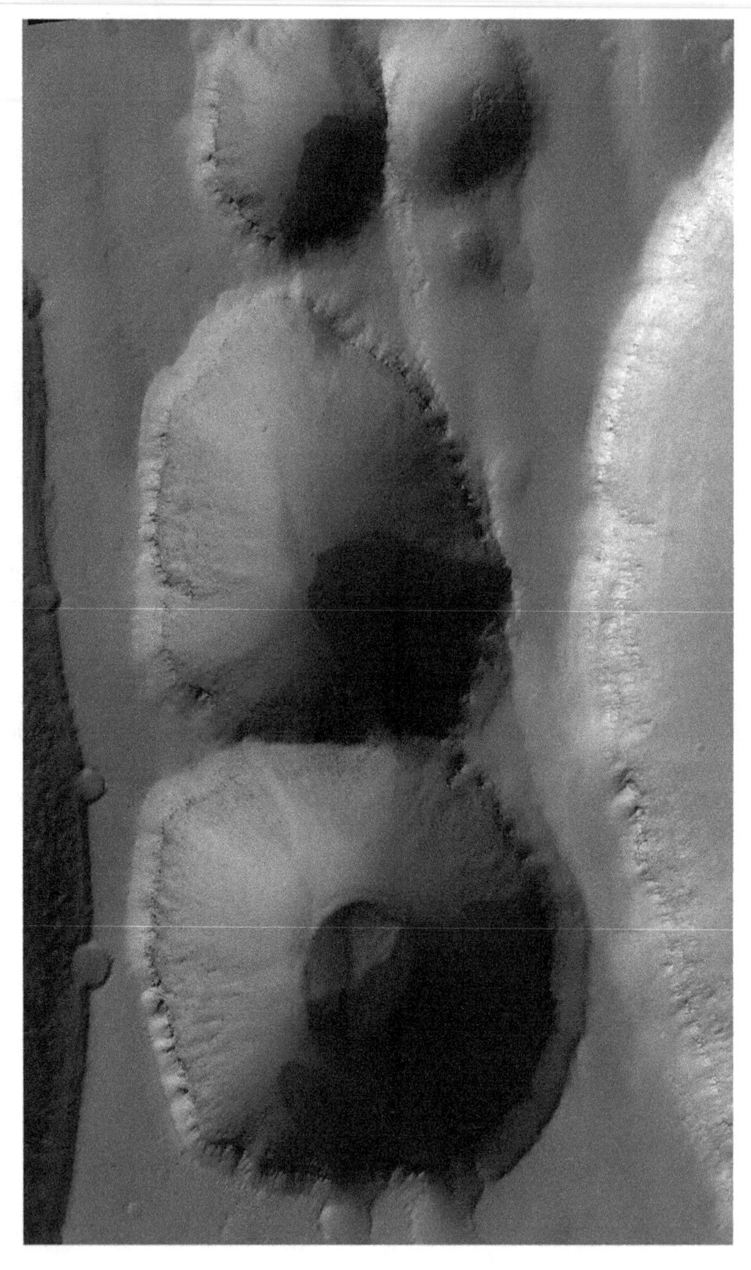

The figure shows a swarm of collapse pits probably created by crust broadening induced by the uplift of the Elysium Mons volcano. North is to the right. Image HiRISE ESP_013144_2075 in the region of Elysium Mons. Scene is about 2 km across. HiRISE (MRO, NASA).

2.1 An Introduction to Mars

Morphometry of Mars, in Brief

The figure at the beginning of the book shows an overall aspect of Mars based on altimetry. Different colors represent the altitude at Mars's surface, measured accurately by the Mars Global Surveyor (MGS) spacecraft based on the return time of a laser beam. On Earth, the sea sets a reference point for the continents (positive altitude, or sea level) and for the oceans (negative altitude, or depth). But since there are no oceans on Mars, the average distance from the center of the planet of 3660 km is used as the reference level, called the datum. The terrain in yellow in the figure corresponds to this elevation. Green, blue, and dark blue represent altitudes below average. The red, the gray, and then the white represent higher altitudes.

We notice an enigmatic feature at once. The overall distribution of the colors shows that Mars is divided into two parts: while the northern hemisphere has few craters, is smooth and has an altitude below the datum, as the frequent blue color indicates, the mountainous south is higher, strongly cratered, and rough. The border between the two parts is well visible on the two maps, but it does not look regular; zigzagging, it does not seem to follow any obvious geometric motif and appears to be strongly affected by massifs along its way. This strange division of the planet into two parts is known as the Martian dichotomy. Figure 2.1 is a black and white version of the same altimeter map with the names of some of Mars' most important regions and structures. Approximately perpendicular to the dichotomous line, there are valleys and canals (only Ares and Kasei are shown in the figure, but there are many others).

In the southern part of the planet, there appear two huge basins more than 1000 km wide, looking like sub-circular depressions. They are the result of the catastrophic impact of asteroids of diameters greater than 100 km. The depth of the bigger one, Hellas Planitia, exceeds 8000 m above the level of the surrounding terrain, comparable to the height of Mount Everest. Argyre Planitia is a little smaller, but still remarkable: about one and a half thousand kilometers in diameter and three to four thousand meters deep. Early colonizers of Mars may be tempted to build a city in the midst of these impact basins, where the artificially introduced air will have pressure greater than average due to the depth of the crater.

The southern part of the planet is divided into a series of Terrae. These southern terrae are saturated by craters due to meteorite impacts that have occurred over four billion years. Terra Cimmeria lies to the east of Hellas Planitia. It is a typical Terra of the southern hemisphere, a very cratered region at high altitude. Arabia Terra and Terra Meridiani are at a slightly lower altitude. Noachis Terra, west of Hellas Planitia and east of Argyre Planitia, includes very old mountains and is named after the most ancient period of Mars, the Noachian. To the west of Argyre Planitia, Aonia Terra and Terra Sirenum have also been distressed by a large number of meteorite impacts.

Before we proceed, we may wonder where the poetic names of the Martian features come from. Of the many nomenclatures proposed for Mars' geography, the one devised by Schiaparelli has persisted, and is still in use today. This type of nomenclature consists of giving a double name to the different morphologies on the planetary surfaces (and not just Mars, but also Venus and Mercury and moons of the

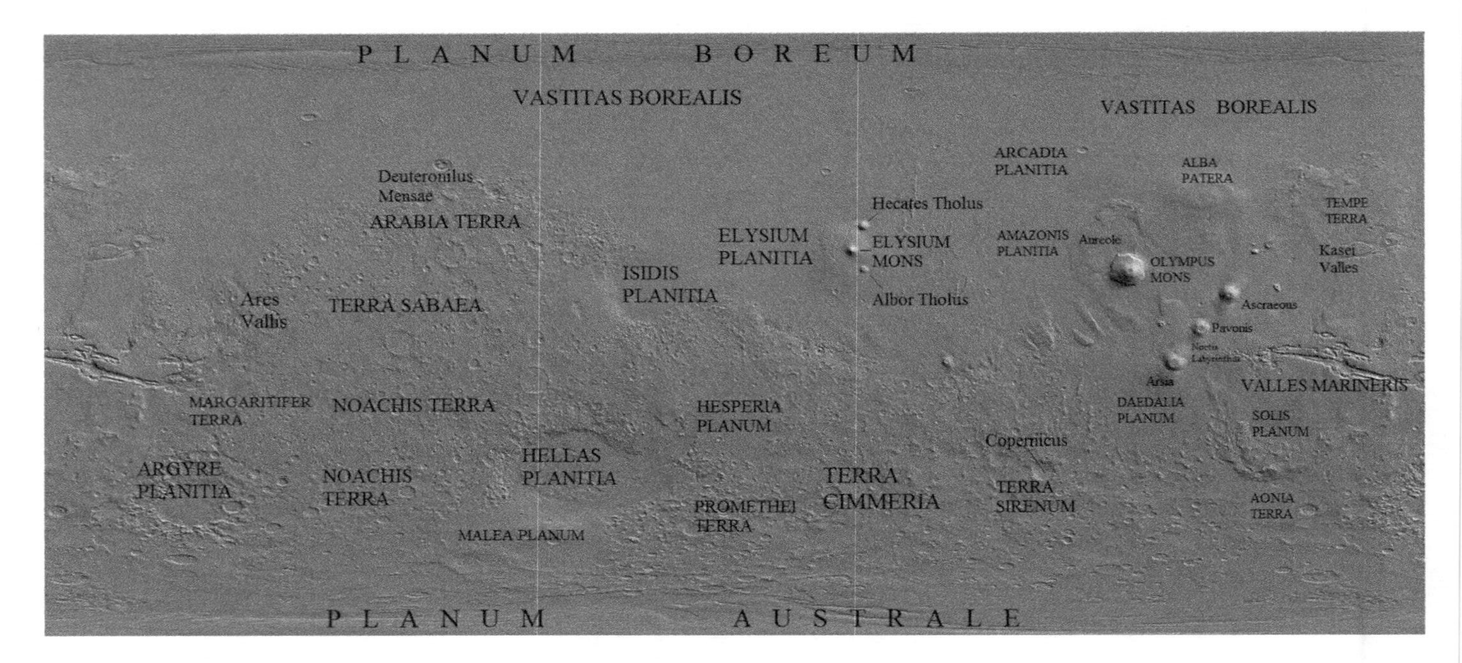

Fig. 2.1 The main units of the Martian physiography. MOLA image with artificial shading effect. MOLA (MSG), NASA

Fig. 2.2 Acknowledged terminology for features on Mars. FVDB

Feature (Designation)	**Description**
Catena, catenae (CA)	Chain of craters
Chaos, chaoses (CH)	Area with broken terrain
Chasma, chasmata (CM)	Long depressions
Collis, colles (CO)	Hills
Corona, coronae (CR)	Ovoid-shaped feature
Crater, craters (AA)	Depression, circular
Dorsum, dorsa (DO)	Ridge
Fossa, fossae (FO)	Depression, long and narrow
Labes, labēs (LA)	Landslide
Labyrinthus, labyrinthi (LB)	Complex system of intersecting fractures
Mensa, mensae (MN)	Highland bound by steep flanks
Mons, montes (MO)	Mountain
Patera, paterae (PE)	Irregular crater-like form
Planitia, planitiae (PL)	Lowland
Planum, plana (PM)	Plateau or highland
Rupes, rupēs (RU)	Cliff
Sinus, sinūs (SI)	Bay
Sulcus, sulci (SU)	Subparallel furrows and ridges
Terra, terrae (TA)	Extensive land mass
Tessera, tesserae (TE)	Tile-like, polygonal terrain
Tholus, tholi (TH)	Small domical mountain or hill
Unda, undae (UN)	Dunes
Vallis, valles (VA)	Valley
Vastitas, vastitates (VS)	Extensive plain

solar system). One name identifies the morphology with a character or place often drawn from Greek–Latin mythology, or from ancient geographic lands. A second name in Latin specifies the type of structure, whether a mountain, a basin, or a fracture, a basin, or a fracture (Fig. 2.2). For example: Amazonis Planitia is a large flat area to the west of Olympus Mons. Because *Planitia* is the Latin name for lowland, Amazonis Planitia means the lowlands of the Amazons, referring to those warrior women who, according to Greek mythology, lived in the Scythian. Terra Cimmeria is a vast land inspired by the ancient name of the Crimea, and so on.

Some names, such as those of structures so large that they can be seen through telescopes, have remained unchanged since Schiaparelli's times. Thus, the names of Noachis, Argyre, Hellas, Elysium, Arabia, Ophir, Thaumasia, and many others can be credited to Schiaparelli. Most, however, were attributed only more recently, after space explorations, while retaining the useful idea of the "double name." The names of structures not visible through telescopes, and even several smaller ones, have been added since the 1960s on the basis of rules ratified by the congresses of the International Astronomical Union. Many features remain unnamed, but this is true even for Earth. The names of most craters (Korolev, Schiaparelli, Gale) are, however, named after people, while flow channels are named after the planet itself in different languages.

Let us go back to the map of Mars. The presence of imposing volcanoes, with their classic conical shape, is apparent. The region with the highest density of volcanoes is that of Tharsis, cut to about half by the equator. Its volcanoes are, in fact, the largest in the entire solar system: Arsia, Pavonis, Ascraeus Montes, Ceranium Tholus, Uranius Patera, and then the two huge Alba Mons (the most voluminous on Mars) and the famous Olympus Mons, the highest volcano in the solar system with its almost 22 km. It is worth noting that the density of that the density of impact craters in this volcanic region is much less than that in the Terrae of the southern area.

To the southwest of Tharsis, there appears one of the most amazing structures on Mars. It is called Valles Marineris, and apparently it is just a canyon, or rather a series of many canyons connected to one another. Deep down to between 7 and 10 km, it is well observable, even by the telescopes orbiting around the Earth. The canyons that we know on Earth have all been dug by a river, but this is not so with Valles Marineris, although the past existence of water and ice in it would seem to explain many of its morphologies. Perhaps Valles Marineris is connected to the volcanoes of Tharsis through deep structures, as evidenced by its main axis, which makes an angle of 60° with the line of volcanoes Arsia–Pavonis–Ascraeus. To the west of Tharsis, there is a second major volcanic area called Elysium. The flows of Elysium Mons travel to the west, running over 1200 km on the northern lowlands. The cause of this attraction to the west was a "hole" over 2000 km in diameter called Utopia Planitia, an impact basin slightly larger than Hellas Planitia. Even on the northern plains, there are huge impact basins, but these are far less profound than those in the south. The first details of the Martian landscape to be described and explained correctly (in 1666) were the two Martian polar caps, at a time when even the terrestrial polar caps were not yet known in Europe. The two caps decrease and increase in volume, depending on the season.

Technical Box 2: Martian Nomenclature

Of the many nomenclatures proposed for Mars' geography, the one proposed by Schiaparelli has stuck, and is still in use today. The terminology proposed by Proctor, based mainly on the names of British scholars, and Flammarion (Continent Herschel, Cassini Land, Kepler Land) have fallen into oblivion. Each structure is designated by two names, one indicating the type of morphology (a list is presented in the table below), together with another name taken from different sources, often Greek–Roman mythology. The charm of

(continued)

Technical Box 2 (continued)
Greek–Latin mythological names, the usefulness of a "double name" identifying the type of geographic structure, and the authority of Schiaparelli have all contributed to this nomenclature's besting of its rival proposals. Today, the names of planetary features are approved by the International Astronomical Union and communicated through the Gazetteer of Planetary Nomenclature. The following table (courtesy of USGS, slightly modified) summarizes most of the Martian features (Fig. 2.2).

The Interior of Mars

The average density of Mars is 3.93 g/cm^3, smaller compared to those of the other terrestrial planets (Mercury 5.44 g/cm^3, Venus 5.25 g/cm^3, Earth 5.52 g/cm^3). This difference is probably related not only to the planet's history, but also to the past dynamics of the whole Solar System. According to the hypothesis of nebula condensation, the less volatile elements condensed towards the center of the Solar System, leaving the lighter ones on the periphery. Mars must therefore have a higher percentage of these lighter elements than Mercury, Venus and Earth, resulting in lesser density. The density of a planet at large pressures is also due to the self-compression of the rocks as a response to the weight of the rocks themselves. So, the more massive a planet is, the greater its gravity, and thus its density. This is confirmed by the low density of the moons of the gaseous planets Jupiter, Saturn, Uranus and Neptune (some of them also contain a lot of ice that lowers the average density).

The dimensional inertia of Mars, measured with precision by the Pathfinder, is 0.365. This number is sensitive to the percentage of planetary mass concentrated in the core. A homogeneous sphere has non-dimensional inertia of 2/5, or 0.4. A ball denser at the edges has a moment of inertia greater than 0.4. The value of Mars, smaller than the constant density value at 0.4, therefore indicates that the density increases toward the core of the planet. This is reasonable, because all planets show a density that increases with depth, which is explained partly by self-compression and also because gravity attracts heavy elements and minerals to the center of the planet, a process called gravitational differentiation. The value of 0.365, intermediate between that of the Earth and that of the Moon, indicates that the processes of differentiation in the interior of Mars have been more efficient than for the Moon, but less efficient compared to the Earth.

It is believed that the core of all terrestrial planets is metallic, composed mainly of a mixture of iron and nickel. The low moment of inertia confirms that the core of Mars, with a probable radius of 1500–2000 km, must consist mostly of iron (average atomic weight 56), with contamination of certain elements such as nickel, cobalt and sulfur. Connected to the internal structure of the Martian core is the mystery of the Martian magnetic field. Or rather its absence, since even Mercury, which is much smaller than Mars, possesses a magnetic field. This problem is related to Mystery no. 1. (Note that there is indeed a magnetic field on Mars, but this is a residual field recorded in the rocks and not something produced today, like on our planet).

The bulk of the planet's mass is contained in the median portion, the mantle, probably about 1500–1800 km thick. It is possibly composed in the deeper part of spinel and majorite (a kind of garnet) and of olivine and garnet at shallower depths. The composition is thus similar to that of the terrestrial interior, with perhaps only a slightly greater percentage of iron. There is, however, a noticeable difference between the Earth's and Mars' interiors. The Earth's mantle is home to strong convection currents, which have a dramatic impact on the surface of our planet, giving rise to the horizontal movements of the tectonic plates, and hence to much of the Earth's geological activity. It is not known whether the Martian mantle is in a convective state, too. It is possible that Mars has a superficial layer close to solidus (or close to melting for a rise in temperature or pressure drop) and that this layer has provided the primary material for the formation of the giant Martian volcanoes. The most superficial layer, the Martian crust, formed as a differentiated gravitational, a bit like the separation of cream from milk. Something like this happened on Earth as well: continents are the distillate of the differentiation of the Earth's mantle. The Martian crust ranges from 25 to 40 km, somewhat slimmer than the Earth's (5 km below the oceans, up to 80 km below the continents).

New Instruments for the Return to Mars

More than twenty years had passed since the success of the Viking missions (1975–1976) when NASA launched two new missions to the Red Planet: Mars Global Surveyor in 1996 and Mars Pathfinder in 1997. Twenty years during which, as we have seen, many missions were attempted by both America and Russia, such as the Phobos missions of 1988–1989, the Mars Observer in 1992, and Mars 96 of 1996.

The Pathfinder reached Mars on July 4, 1997. Renewed scientific interest in Mars after a lull of twenty years required rethinking and new strategies. That is why the mission was carried out with such scarce scientific equipment, mostly provided by Europe. The mission also required the testing of new landing techniques with parachutes, balloons and rockets. Newly-designed instruments were also tested. APXS (Alpha Proton X-ray spectrometer) could reveal chemical elements based on the diffusion of alpha particles emitted by a source. The stereo camera in the orbit allowed for new photos of Mars' surface. Studies on the atmosphere and its diurnal variations were reestablished, and determination of the moment of inertia was undertaken.

But the true symbol of the new era of exploration of Mars was the Sojourner, a rover the size of a toy. The Sojourner was the first rover on Mars, although it was fully controlled from Earth, and was therefore very limited in its movements, since it could only leave the lander for a distance of a few tens of meters, proceeding very slowly. Shipped to an area where there must have once been huge floods, it found a silica content with greater similarity to andesites than basalts, and discovered the sandy vortices in the atmosphere known as dust devils. The new determination of the moment of inertia based on comparison between the new mission and the old Viking data showed that Mars has a metallic core of radius between 1300 to 2400 km.

The other mission, that of the Mars Global Surveyor (launched before Pathfinder but not rendered operational until later), was primarily intended to study the entire topography of the planet on a large scale. The data from this spacecraft is still widely used for research on Mars' morphology and geophysics (the 25,000 images from the Vikings, while still useful, have been replaced by those of subsequent missions, while those of the Mariner are now only of historical interest). MGS carried, among other things, the MOC optic camera to provide an overview of images with better resolution and the Mars Orbiter Laser Altimeter (MOLA) to accurately measure the topography of Mars, of fundamental importance for each subsequent study of the morphology of Mars, while the Thermal Emission Spectrometer (TES) provided analysis of the mineralogical composition, starting from the spectrum of reflection. Additionally, another instrument, the MAG/ER, a magnetometer and a reflector, allowed for the first measurement of the Martian magnetic field.

The Magnetic Field

When a spacecraft approaches Mars to enter a closed orbit, it has to slow down its speed with respect to the planet. If it were not held back, the probe could never go into orbit around the planet, because its different velocity would force it away after only a brief greeting to the planet. This is indeed what happens during the so-called flyby missions, such as the recent, breathtaking passage of the New Horizons mission to Pluto. In this kind of flyby mission, space engineers need to exploit the short duration when the probe, travelling at several tens of km per second, meets the planet travelling along a different direction dictated by its orbit. In this frantic situation, lasting, at most, a few hours, images need to be taken at the highest possible rate before being transmitted back to Earth. In contrast to a flyby, a rendezvous mission requires equating the velocity between the spacecraft and the planet. This is precisely the aim of aerobraking. In 1997, the aerobraking technique was attempted with the Mars Global Surveyor (MGS). The spacecraft was placed on a strongly elliptic orbit, so that the periaster was within the Martian atmosphere. An atmosphere that, though much thinner than that of the Earth, still exerts a braking effect on a spacecraft that moves at several kilometers per second. The spacecraft's cross-section (and therefore the force opposed to the atmosphere) was increased with solar panels, which, however, risked breakage during the one and a half year period of the maneuver's duration. As the spacecraft revolved around the planet thousands of times at a distance of about 400 km from the Martian surface, it became possible to measure the magnetic field at that distance using the MGS MAG/ER magnetometer aboard.

Since the Mariner 4, it had been known that the Martian magnetic field, if any, had to be very small, but a precise measure had never been made. The weak field was confirmed: only 400 nT (nanotesla) at the equator, a value much smaller than the 30,000 nT of the Earth's magnetic field. Such low values for Mars indicate that the field is not generated continuously. What was measured was only a residual magnetic field, a kind of fossilized magnetism in the Martian rocks. Moreover, the field was found to be stronger in the southern part, where it was strangely distributed in strips (Fig. 2.3).

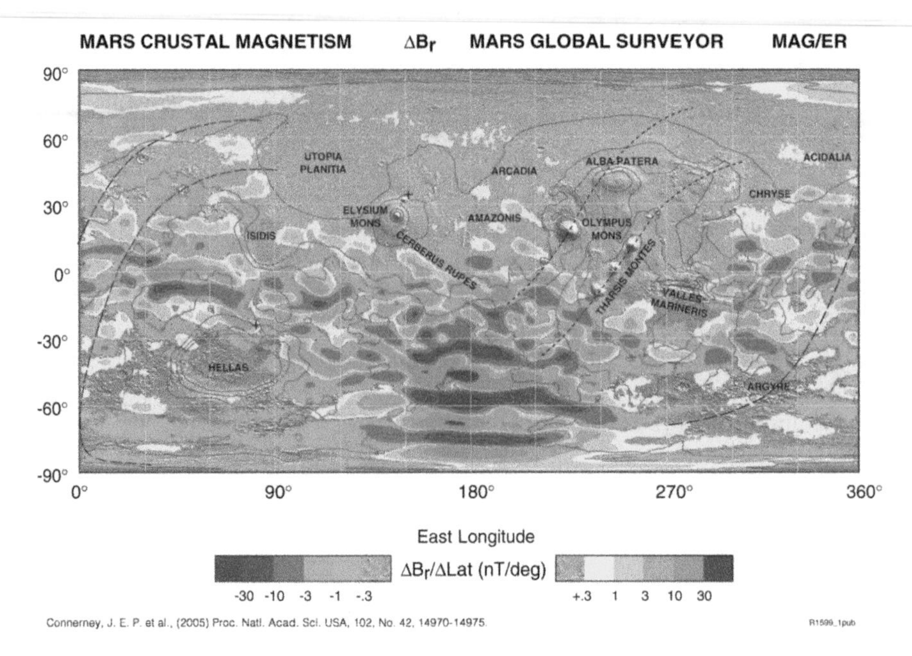

Fig. 2.3 A map showing the radial component of the magnetic field of Mars at 400 m altitude. Both the magnetometer and the electron reflectometer of the MAG/ER instrument aboard MGS are used in conjunction. The blue stripes correspond to a field directed toward the planet's surface, while along the red stripes, the field is pointing outward. The field data are superimposed onto a MOLA shaded relief map. Note that the field is stronger in the southern part of the planet. NASA (MGS)

Most people know the terrestrial magnetic field only from the compass. For a planetary scientist, the magnetic fields of the planets reveal very important details about both the core of the planet and its thermal history, and also about the nature of the rocks on the surface. To clarify how the magnetic field originates in a planet, consider the Earth. Because of the heat flow directed from the interior toward the surface, both the Earth's mantle and its core are home to convective movements similar to the rise of hot air above a candle. The vertical motions in the mantle are slow, as they occur in solid rock. Nevertheless, they have a very important effect on the characteristics of our planet's geological activity. Vertical movements translate into the horizontal movements of the lithospheric plates at the rate of a few centimeters per year, resulting in continental drift. Convection motion in the outer core is responsible for origination of planetary magnetic fields. This is because the medium generating the field must be both metallic and in a liquid state. The absence of the magnetic field would therefore indicate a modest role of convection in this outer core. The reason why both metallicity and a liquid state are needed is that the magnetic field is originated by internal currents by means of a self-reinforcing dynamo mechanism.

In short, the mechanism is as follows. Because the core is in liquid form, convective movements are fast and involve a liquid metal, and are thus conductive of electricity. Coriolis's force, the same one that, owing to Earth's spinning state, transforms the masses of air in the atmosphere into huge vortices like the great

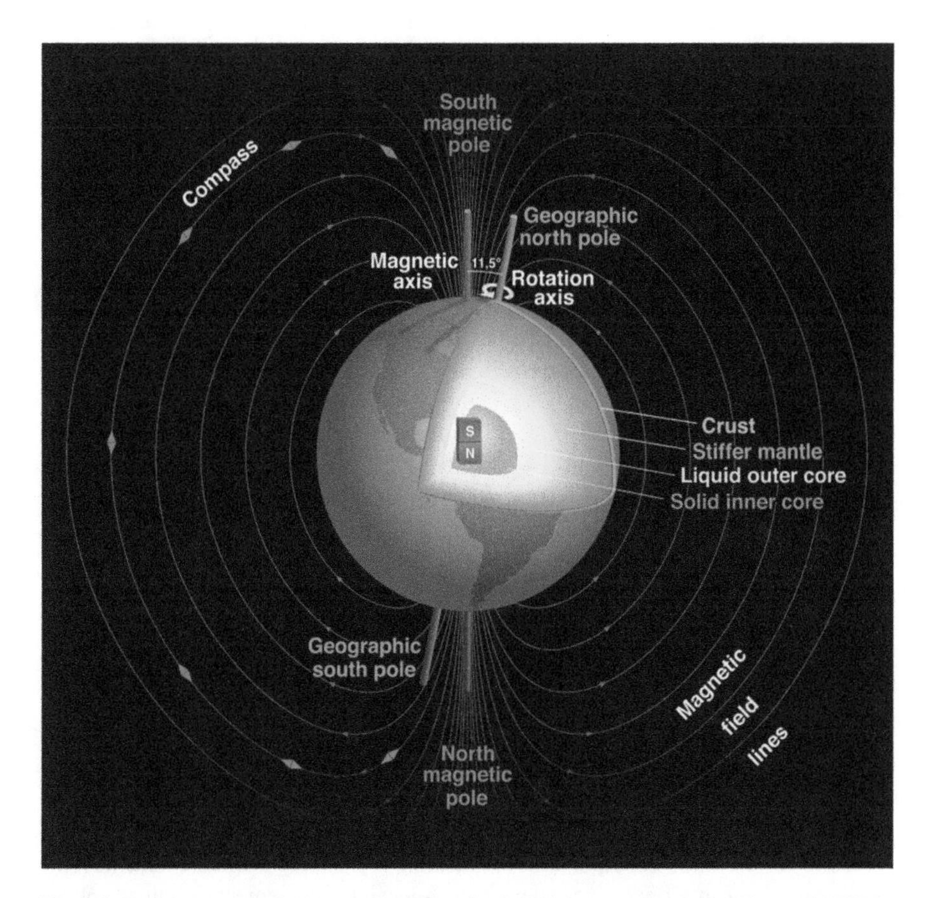

Fig. 2.4 Earth's magnetic field, approximately a dipole field, is created in the liquid core. FOTOLIA

anticyclones or hurricanes, is of crucial importance for the generation of the magnetic field. This is because the Coriolis force thrusts the fluid into a state of movement, forming trajectories of cylindrical shape aligned with the axis of rotation. In this way, the magnetic field grows, because the movement of the fluid in turn reinforces the magnetic field. In fact, a moving current generates a magnetic field according to well-established Maxwell's electromagnetism laws. From these complicated movements in the core, a magnetic field is created similar to the one of a dipole, that is, the one that would be generated by a hypothetical, huge cylindrical magnet piercing through the length of the Earth from North to South (Fig. 2.4). If convective movements cease, the magnetic field does not die immediately, but rather decays within geologically short periods of time.

When a terrestrial magmatic rock solidifies, the crystals of ferromagnetic minerals (minerals containing iron such as magnetite and pyrrotite) gain a magnetization aligned with the magnetic dipole field present at solidification. Once acquired, this orientation continues to be preserved after the field dies. On Earth, examining an ancient magmatic rock tells us about the history of the Earth's magnetic field. On both sides of the Earth's oceanic ridges, there are lines of magnetic polarity that form parallel stripes. Because each stripe has magnetic polarity opposite to that of the adjacent ones, stripes indicate a reversal of the magnetic field as the magma gusted out of the dorsal. These magnetic stripes allowed for a general acceptance of plate tectonics. Moreover, the direction of the magnetic field tells us about the positions the magnetic pole occupied in the past, and how the continents have moved since then. If the dipolar magnetic field dies, the residual magnetic field remains attached to the rocks. Lots of valuable information, then. But let us turn our attention back to Mars.

The MGS mission on Mars, and later the MRO, observed only the fossil magnetic field on the Martian surface (Fig. 2.3). Field distribution on the surface is strangely similar to the magnetic stripes on Earth mentioned earlier. Also, on Mars, the stripes shown in blue and red have opposite magnetic polarity. Towards the north, the magnetic field becomes much smaller, and indeed disappears in the volcanic area of Tharsis. The field is also zero in correspondence with the giant impact basins of Hellas, Argyre and Utopia Planitiae and in the northern part of the planet, dominated by the vast lowlands. The most intense magnetization is observed in the very ancient regions of Terra Cimmeria and Terra Sirenum.

Because the field remains strongest in the southern areas of Mars, which are also the oldest, the planet must have been capable of generating a magnetic field with the mechanism of the dynamo, but this primary field is now dead, and only the weak field in the rocks is measured. According to most specialists, the field survived only 100–500 million years after the formation of the planet.

FIRST MYSTERY: What Killed the Magnetic Field?

The enigmatic characteristics of Mars' magnetic field are therefore threefold: why and when did it disappear? Why does the weak residual magnetic field reside mostly in the southern hemisphere? How did the peculiar striped pattern originate?

We recall the three conditions for maintaining a planetary magnetic field with the dynamo mechanism. Convective and rapid heat-generated movements within the planet must occur; convection should partly include a metallic, electrically conductive fluid; and finally, the planet must rotate at sufficient speed to divert convective motions with the Coriolis effect. On Earth, the only place where all three of these conditions can be met is the core, composed mainly of electricity-conducting iron. While the inner part of the core is solid, its outer part, due to lower pressure, is in liquid form. As for the third condition, the Earth rotates on its axis quite quickly. If the dynamo mechanism for the magnetic field has disappeared from Mars (although there was a time when it was active, as the fossil magnetic field shows), then all three conditions were met at the beginning of the Red Planet's story, but now, at least one of the three conditions is no longer valid.

Venus has no magnetic field, although it has many characteristics similar to Earth, including comparable size (it is considered the Earth's twin planet, despite having a temperature on the surface capable of melting lead). Venus must also have an external iron and liquid core in a state of thermal convection. But due to tidal effects with the Sun, with time, Venus slowed its rotation speed around its axis. Today, it has a rotation period of 243 days, corresponding to an angular velocity of only 0.4% that of Earth's. The absence of a magnetic field on Venus can therefore be explained by the slow rotation around its axis. Mars has a rotational speed comparable to that of the Earth, so we cannot invoke the same explanation that applies to Venus.

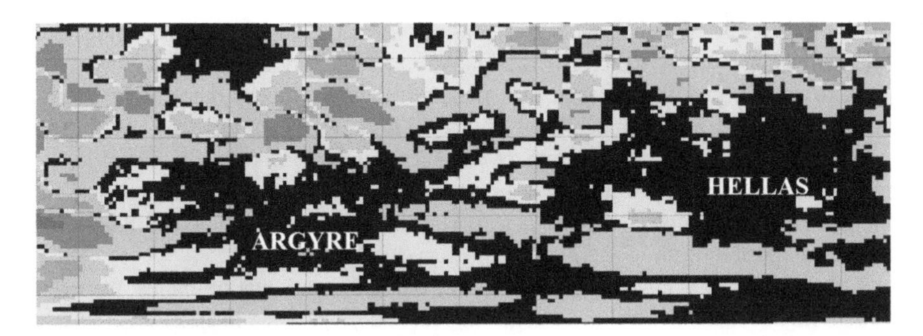

Fig. 2.5 The residual magnetic field is absent in correspondence with the giant Hellas and Argyre Planitiae. Argyre and Hellas are the products of the early impact of two asteroids about 100 km in diameter or more. The energy and warmth during the explosion that followed the impact was so huge that rock was thrown hundreds of kilometers away, and the one that remained at the base of Hellas was heated beyond the Curie temperature, at which iron in the iron-bearing minerals of the basalts loses its magnetization. As a consequence, the primordial magnetic field was stripped away in these areas. NASA (MGS)

We know, however, that the liquid iron of the outer core of the Earth at the boundary with the inner core is solidifying. As a result, the solid internal core is constantly rising at the expense of the outer core at a rate of about 1 cm per century. The magnetic field is also affected by the decrease in available liquid iron in the outer core, causing a weakening of the dynamo mechanism. For the Earth, at least a billion years will pass before the magnetic field will be sensibly weaker than today. Some researchers think that this process has already concluded for Mars, and now the core is entirely solid. In fact, Mars is a smaller planet than Earth, with a core volume just one tenth that of the Earth. However, Sun-induced tide measures seem to show that part of the outer core is still in liquid form.[1] Thus, we cannot conclude that the core of Mars is completely solidified today.

The calculations show that after the core has completely solidified, the magnetic field of a planet like Earth or Mars would die in just ten to twenty thousand years. Thus, the age of the fossil magnetic field on the Martian rocks gives us information as to the last period when the magnetic field generation in the core was active. It is interesting to note, on the basis of Figs. 2.3 and 2.5, that the magnetic field is missing at the two great impact basins of Hellas and Argyre Planitiae in the south of the planet. It seems that the fall of the two asteroids that created such basins wiped out the field instantly. The explanation is that the fall of a large meteorite releases a very high energy capable of heating the rock in the impact region to a thousand degrees (perhaps even more so in the case of such large impacts). A magnetized material such as a rock or metal loses its magnetization above a critical temperature, known as Curie's temperature, which, for iron, is at 768 °C. The impacts of Hellas and Argyre therefore caused a real rearrangement, both mechanical and thermal, of the Martian crust, erasing the magnetic field from the superficial rocks. This implies that at the time of the impacts, the dynamo mechanism had already disappeared, otherwise the field would have been reformed in

[1]Konopliv, Alex S., et al. "Mars high resolution gravity fields from MRO, Mars seasonal gravity, and other dynamical parameters." *Icarus* 211.1 (2011): 401–428.

the cavity left after the impacts. Both Hellas and Argyre Planitiae date from 3.8 to 4.1 billion years ago, the so-called late heavy bombardment period.

How to interpret the absence of a magnetic field in the northern hemisphere? According to most researchers, the magnetic field was once also present in the northern lowlands. But a huge impact (or a series of impacts) hit the planet to the north, warming it beyond the Curie point. Here, the mystery of the magnetic field blends with that of global dichotomy (the Third Mystery), i.e., the diversity between the north and the south of the planet.

Other researchers have suggested that the paleo-dynamo that originated the magnetic field in the Noachian (up to 4 billion years ago) was asymmetric. According to this hypothesis, investigated through computer simulations, the heat flow to the south of the planet was much more intense than in the north. The result of the calculations shows that only the part of the outer core facing one hemisphere of the planet was subjected to convection in the outer core, and thus could create a magnetic field. This hypothesis, however, does not explain the reason for such deep asymmetry in a planet that, due to its high gravity, should have a fairly uniform interior mass distribution, nor does it explain the characteristic magnetized strip distribution.

Perhaps the residual lines of the field were created during the expansion of the Martian crust in a timid phase, subsequently aborted, of plate tectonics? On Earth, the stripes are observed in the Earth's oceanic crust, not on the continents: on the most recent rocks, not the oldest. The southern hemisphere, where the magnetic field is stronger and the stripes are more evident, has more affinity with the continental crust than with the Earth's oceanic crust. It is also possible that these enigmatic stripes have had similar origins, but with substantial differences.

Other researchers have suggested that Mars' magnetic abnormalities are the remnants of the drifts of hot spots at the time the magnetic field was present.[2] Such drifts would be caused tidally by hypothetical moons of Mars, now gone (perhaps as a result of having collided with Mars, producing the large impact basins).

The disappearance of the Martian magnetic field was not painful for the Red Planet. Deprived of a magnetosphere, the surface remained exposed to cosmic rays, which contributed to the formation of the regolith and probably to the dispersion of the Martian atmosphere.

2.2 The Story of Mars in Brief

Early Mars

The solar system has remarkable regularity. The planets move approximately along the same plane and rotate around the Sun in the same direction in almost circular orbits, as do their satellites. This configuration cannot be obtained from celestial

[2]Kobayashi, Daisuke, and Kenneth F. Sprenke. "Lithospheric drift on early Mars: evidence in the magnetic field." *Icarus* 210.1 (2010): 37–42.

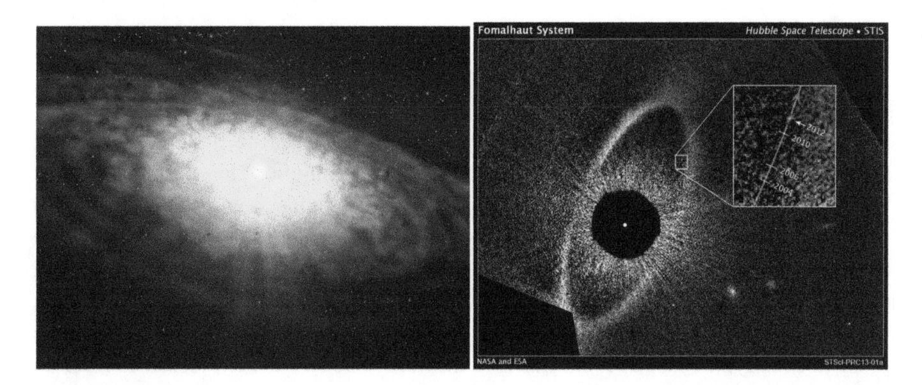

Fig. 2.6 Left: An artistic conception of the formation of the solar system. Right: The planet Fomalhaut b (showed and enlarged in the box) is orbiting its star Fomalhaut in a highly elliptical orbit in an environment still full of debris. The central "hole" is due to the chromatograph screening the strong stellar light. Left: image 118292360 FOTOLIA/Mopic, right: NASA (Hubble Space Telescope)

bodies that have fallen randomly into the solar gravitational field. The philosopher Immanuel Kant (1724–1804) realized that the solar system was derived from the collapse of a proto-nebula subjected to its own gravity. According to Kant, the cloud began to spin around an axis (why and how a rotation axis could form from the random movement of gas and dust was not known at the time), and in the center, the Sun was formed. According to this hypothesis, our star and its planets formed at the same time, something that was not so obvious at the time of Kant, but which is today considered perfectly valid. The mathematical physicist Pierre Simon de Laplace (1749–1827) was the first to take this hypothesis very seriously and rework it. The planets would result from the aggregation of gases and dust, which would continue to collide during the formation of the solar system. Today, the accepted model for the formation of the solar system stems from the Kant-Laplace concept (Fig. 2.6, left). Obviously, it has evolved a long way beyond the original hypothesis, and modern studies rest on elaborate numerical simulations, even though some facts are still unclear, such as the too slow speed of Sun rotation (according to theory, the Sun should rotate around its axis in a few seconds, while its rotation period is, in fact, almost a month) and the details of the aggregation process of planetesimals. Confirmation derives from the relatively recent observation of planetary systems in formation around other stars (Fig. 2.6, right).

Once aggregated, each planet followed its own evolution, which, though initiated by a common process, was, however, independent. One of the major differences between the planets was their distance from the Sun. The planets closest to the Sun (Mercury, Venus, Earth and Mars) were depleted of gas, compared to the gigantic gaseous planets, farther away. As a result, the four terrestrial planets today have a solid surface. Another important difference was the mass at which the process of aggregation was formed. The two largest terrestrial planets, Venus and Earth, have a fairly thick atmosphere. The other two, Mercury and Mars, smaller, and therefore endowed with a lower gravitational field, were incapable of retaining much gas.

The Periods of Mars History

The history we know best is, of course, that of our own planet; a complicated story that can be roughly distinguished in three phases. During the first phase, lasting about two billion years, Earth's temperature was much higher than today due to the heat released by the planet in contraction, meteorite impacts, and the contribution from the decay of radioactive nuclei, which was also greater than today. All of these phenomena also take place today, but at a much lower rate. The second period, another roughly two billion years, saw the proliferation of life in primitive forms. Oxygen increased and the atmospheric composition became closer to that of the present. Finally, the last five–six hundred million years saw the explosion of life into the forms we know today. This is the Phanerozoic eon, that is, of "visible life," which, in turn, is distinguished by eras, periods and a myriad of stratigraphic sub-units in which we find most of the known fossils that can be seen in museums.

Mars' story is far less known for three obvious reasons. First, we have limited access to its surface. No geologist has ever held a Martian rock, with the exception of some meteorites from the Red Planet. In addition, the photographic, altimetry and radar images have only been available for a very few years. Thirdly, it does not seem that Mars has ever hosted life, which is an awesome marker of a planet's vicissitudes. And even if one day, fossils are found on Mars, they will likely bear witness to small forms of primitive life, similar to those of the early stage in terrestrial history, and nothing of the profusion of life that has characterized our planet. On Earth, geological dating techniques are multiple: absolute dating, such as decaying some radioactive nucleic acids, and fossil analysis. Because of these uncertainties, it is obvious that the story of Mars cannot be known with the same degree of precision with which we know the story of our planet today (this is even more true for the last half billion years, for which the Earth's chronology is also based on fossils, not to mention the many dating techniques available for more recent finds).

Yet, it is impossible to reconstruct the geological history of Mars without dating important events. For the time being, there are only two possible methods available for dating Martian terrains: the subsequent overlap of formations, structures, erosive phenomena or geological deposits, and crater-counting techniques. The first method is very simple: if a certain geological structure, such as a lava stream, a landslide or a sedimentary deposit, rests on another formation, then it must be the latest of the two (Fig. 2.7). This method has two obvious limitations. Not only does it provide no absolute age estimate (nor the age difference between the two structures), it is also based on identifying boundaries between the various structures, which are often ambiguous. Figure 2.8 shows a more complex example.

More reliable but not of immediate application is the second criterion, based on a simple principle. Rocky bodies like meteorites[3] and comets fall with some frequency

[3]To be more precise, solid bodies entering the Earth's atmosphere and giving origin to a shooting star, or a more luminous (and often noisy) bolide, are called "meteoroids." Note the terminology, putting emphasis on our own atmosphere, rather than on the solid body itself, and dating back to a time when shooting stars had no explanation. If the solid body does not entirely wear out in the

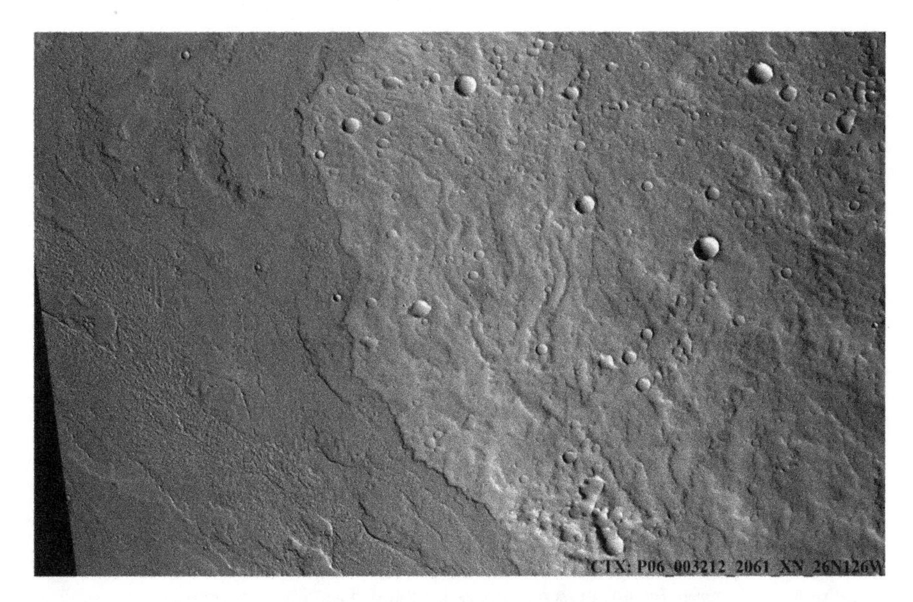

Fig. 2.7 The lava flow on the left from Olympus Mons is superimposed onto the flows from Alba Mons on the right, and is therefore younger. This is evident from the "freshness" of the more recent flows, and is confirmed by the smaller number and size of the impact craters. CTX image. CTX (MRO), NASA

on all planets. The frequency was greater in antiquity than it is today, because most of the solid material in the solar system is pristine and is not replenished. Everybody has experienced shooting stars, a flash in the atmosphere resulting from tiny particles that disintegrate in a glorious final flash, after having spent billions of boring years waiting in space. A large meteorite resists atmospheric ablation until it hits the ground and normally creates an impact crater whose size depends on speed and mass, as well as the characteristics of the planetary surface (kind of rock, gravity field, presence of water or ice).

The more a particular surface has been exposed to meteorite fall, the more craters it has, and these craters will be, on average, larger, because most of the largest meteorites and asteroids fall within one billion years of the formation of the solar system. Thus, it is possible to date the surface based on the statistical analysis of such impact craters.[4] Based on this technique (Technical Box 3), four periods have been distinguished in Mars' history: an initial period of accretion and differentiation (which can be called the Pre-Noachian) from which no rocks have been preserved, followed by three long periods called the Noachian, Hesperian, and Amazonian (Fig. 2.8).

---

atmosphere, once found on the ground, it is called a "fall." A "find" is a rock fallen from space that has not, however, been seen falling. Finds and falls are collectively called "meteorites." In the book, we use "meteorite" to refer to any solid body fallen from space, also extending this name to objects travelling through the atmosphere.

[4]Techniques have been elaborated by William Hartmann, Gerard Neukum, and others.

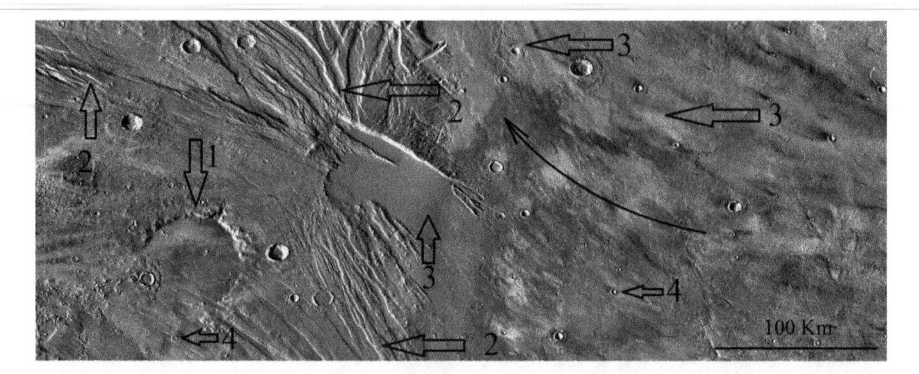

Fig. 2.8 Another example of the dating of Martian terrains in the eastern part of the Acheron ridge. After formation of the ridge, large asteroid impacts created enormous impact basins, such as the one indicated by the number "1." The fact that craters like the one indicated were later opened by the fractures on the left (indicated by the number "2") suggests that the fractures opened up after the craters. Subsequently, the lava flows from the Alba Mons volcano to the east of the figure, well outside the scene, have filled many of these fractures. They must then be successive to them. Note that the lava flows, indicated by the number "3," have deviated north (as indicated by the curved arrow) because of the influence of Acheron. Finally, other craters (some indicated as "4") were formed after all of these events. THEMIS (Mars Odyssey), NASA, NASA/JPL-Caltech/Arizona State University

Pre-Noachian (From the Planet's Formation 4.6 Billion Years Ago to About 4.1 Billion Years Ago). Duration: 500 Million Years

This period goes from the formation of Mars and other planets of the solar system, about 4.6 billion years ago, to the next period, the Noachian. Initially, radiogenic heat, along with heat developed by gravitational differentiation and meteorite impacts, kept the surface of the planet fused with ocean magma. The Martian crust formed within the first 100–200 million years. During the pre-Noachian, Mars had a magnetic field, which had already been turned off by the end of the period. Once the crust was formed, the large presence of debris in the solar system created many impact craters. The formation of Martian dichotomy dates to the pre-Noachian period. Towards the end of the period, planets experienced an increase in the number of impacts, not only on Mars, but on all celestial bodies, possibly caused by the gravitational instability of Oort's cloud, the mass of debris that orbits about a light year from the Sun. Called late heavy bombardment, this episodic bombardment was probably responsible for the excavation of the enormous impact basins of Hellas and Argyre Planitiae. The formation of these two impact basins marks the beginning of the Noachian. On Earth, the differentiation and formation of the continental crust was completed.

Fig. 2.9 Noachis Terra. Part of Viking 623A67. Courtesy of NASA/JPL. Viking (NASA)

Noachian (From 4.1 to About 3.7 Billion Years Ago). Duration: 400 Million Years

Numerous and sometimes gigantic impacts continued, and volcanic activity was represented by enormous basaltic expansions and perhaps ultrabasic lava, as in the rims of Hellas Planitia. The atmosphere was denser and the temperatures higher. Liquid water was present on the surface of the planet, organized in a hydrological cycle, as demonstrated by the presence of torrent beds of this age and the probable existence of an ocean in the northern part of Mars. It is likely that the formation of the volcanoes of Tharsis (Olympus Mons, Alba Mons, and others) dates back to the Noachian. At the same time, on the Moon, the Fra Mauro crater was shaped (4 billion years ago), as well as the oldest volcanic provinces.

The period is named after Noachis Terra, one of the oldest and most cratered areas in the southern region of the planet (Fig. 2.9). During the Noachian, the frequency of crater formation decreased. The terrains dating back to the beginning of the Noachian show about one crater each larger than 1 km in diameter per km^2. The terrains at the end of the Noachian have "only" one crater larger than 1 km for every 80 km^2.

Hesperian (From 3.7 to 3 Billion Years Ago). Duration: 700 Million Years

The Hesperian is considered the principal period of Mars' volcanic activity. The volcanism, which may have had its origin in the Noachian, continues in the northern part of the planet, and especially in the region of Tharsis. Most of the activity of the tallest volcano, Olympus Mons, is considered to date back to this period, even though it is impossible to determine previous activity, since the latest lava

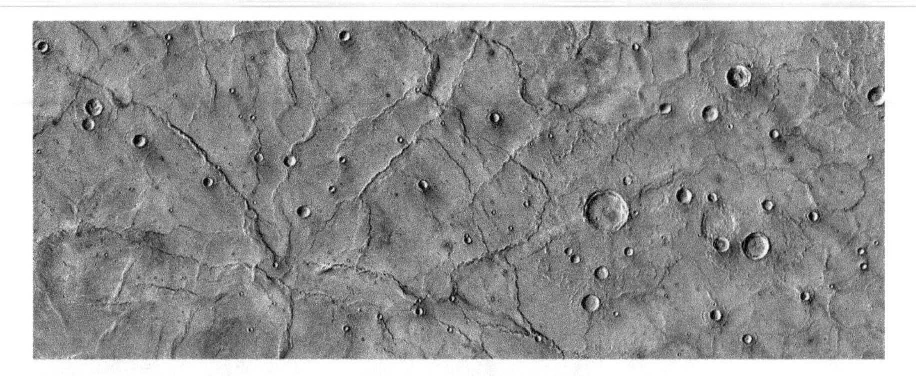

Fig. 2.10 Hesperia Planum (east of Hellas Planitia) lends its name to the Hesperian period. THEMIS image 20.6^{0} ^{N}, 114.8^{0} ^{W}. Image 600 km across. Amazonian (from 3 billion years ago to today). Duration: 3 billion years. THEMIS (Mars Odyssey), NASA

expansions and the enormous growth of these volcanoes masked any previous structures. Also, the Elysium volcanoes may date back to this era.

The period is named after Hesperia Planum, an area blanketed by immense lava expansions (Fig. 2.10). Possible convection developed in the Martian mantle, albeit less energetic than in contemporary Earth. Valles Marineris was also formed in the Hesperian as a result of fractures in the Tharsis area. The mean temperature decreased and snow fields were formed at the top peaks and the poles. There was a possible presence of permafrost on almost the entire surface of the planet, while part of the carbon dioxide in the atmosphere was seized to form carbonates. The Hesperian age was also the period of formation of the major outflow channels that run from the mountainous areas of the south to the north. Meteorite frequency decreased radically compared to the Noachian. On the Moon, the lava fields of the great seas (Tranquility, Fertility) corresponded to the early Hesperian of Mars (3.5 billion years ago). The terrains at the end of the Hesperian had 1.5 craters wider than 1 km every 1000 km^2.

Amazonian (From 3 Billion Years Ago to Recent. Duration: 3 Billion Years)

About three billion years ago, the amount of free water on Mars' surface began to decline and glacial structures became prevalent over outflow channels. The volcanoes continued their eruptive activity, but it seems to have been less intense. Due to the effect of cosmic rays on a planet devoid of a magnetic field, the atmosphere of Mars was disappearing.

The period is named after Amazonis Planitia, a flat area west of Olympus Mons. The fractures of Tharsis and Valles Marineris deepened with the development of gravitational instability and landslides. The climate became colder and drier than the previous periods, and towards the end of the period, the water in liquid form may have already disappeared completely.

Late Amazonian terrains 600 million years old are characterized by approximately three craters per 10,000 km^2.

Technical Box 3: Dating the Martian Terrains

An estimate of how much time we must wait for the bus can be made instinctively based on the number of people at the stop. The presence of many people indicates that the bus is about to arrive, but if we are the only ones at the bus stop, the bus has just passed! A conceptually similar criterion, but, of course, one much more rigorous, is that used to estimate the ages of Martian surfaces. The method is based on the number of craters on a given surface. Intuitively, it is simple: the presence of many craters indicates a land that has remained under the shower of meteorite impacts for a long time. Contrastingly, an area with craters of smaller number and size must be younger.

If the concept is simple, the application to the study of the ages of planetary surfaces (especially the Moon and Mars) has required remarkable efforts by specialists, and here, we can only make a brief summary of what is commonly done to date the Martian surfaces. Let us start with the Moon's terrain. Contrary to Mars, for the Moon we have absolute radiometric dating at our disposal. This is achieved by studying the decay of the radioactive isotopes contained inside the rocks brought to Earth by the Apollo missions. Once the absolute age was determined in this way for some specific locations visited by the Apollos, scientists placed these ages in relation to the frequency of craters in the given area. This comparison resulted in a relationship between the frequency of craters on a certain surface and the radiometric age of that surface. Based on Apollo 16 and 17s isotopic analyses, for example, the Tycho crater was found to be quite young (180 million years). Lunar surfaces of this age typically show a crater larger than 1 km in diameter per 10,000 km^2. The dark areas of the Moon (Maria –meaning seas in Latin) are composed of dark basalt and are much older. On average, a few craters larger than 1 km are counted every thousand square kilometers, and their radiometric age is, in fact, about 3–3.5 billion years. Finally, the Terrae, the lightest parts of the Moon made up of a light rock called anorthosite, are even more ancient: they show an age of more than 3.5–4 billion years and at least one crater larger than 1 km in diameter every hundred square kilometers, but often ten times more. The lunar dates may explain the evolution of the Moon. In short, at the beginning the oldest, anorthosite skin was formed uniformly around our Moon. The continuous fall of meteorites and large asteroids pierced the anorthosite skin, allowing the darker and heavier basalt to gush out, creating the black Maria. By extrapolating data to areas not visited by lunar missions, we can then date all lunar surfaces.

On Mars, we do not have the option of engaging in any radiometric dating (except for a few meteorites), at least until we can send a mission that returns with some samples. However, it is believed that with appropriate calibrations, one can recreate the relationship between the age of a surface and the number

(continued)

Technical Box 3 (continued)
of craters of diameter greater than 1 km. To this end, it is necessary to take into account the possibility that the asteroid flow to the surface was different, with different asteroid impact velocities and different gravity and soil characteristics. These factors do not seem to have dramatically changed the number and distribution of impacts, and hence the extrapolated curve for Mars is quite similar (Fig. 2.11).

In addition to the cumulative curve for craters greater than 1 km, we can have that relative to 2 km, 3 km, or any other measure. In this way, it is possible to calculate calibration curves that are used for the more precise calculation of the age of a surface. The procedure is then:

(1) Establish the region the age of which you want to determine and collect the best images of that area.
(2) Using a polygon, delimit the desired area, excluding areas where craters may have been covered by dunes or other more recent deposits.
(3) For each crater within the area, measure the diameter of the crater and list the values in a file. Correction procedures may be adopted to deal with irregular craters.
(4) Create a frequency curve,[5] which will have an appearance such as that shown in Fig. 2.11. This is a cumulative curve, indicating the best fit for the number of craters with a diameter greater than that indicated.
(5) Compare with the calibration curves, allowing for estimation of the age of the surface.

There are, however, some complications. One problem is that the area required for these counts should be wide for precise age determination, and this implies a large number of craters to be determined. There may also be secondary craters around, coming from larger impacts in the neighborhood, which showered the area under study with smaller impactors. Secondary craters, sometimes recognizable by their alignments, must, of course, be excluded from the statistics. There are also crater clusters, created by the explosion of a meteoroid in the Martian atmosphere. In addition to these caveats, note that, as Fig. 2.11 (left) shows, the continuous lines provide ideal ages, but the real counts typically cross them, showing that age is not univocally determined. These facts and the apparent saturation in cumulative curve frequency for small craters (visible as the flattening out of the curve to the left of Fig. 2.11) indicate the effect of resurfacing, i.e., covering of a small crater by recent deposits and biased exclusion of smaller craters. In this case,

(continued)

[5]The software used for this example, CraterStats, was developed at the University of Berlin, see G. Michael,ftp://pdsimage2.wr.usgs.gov/pub/pigpen/tutorials/FreieUni_Workshop2012/4_craterStats.pdf.

Technical Box 3 (continued)
the smaller craters should be excluded, and the estimated age is about 1 Ga (one billion years). Fig. 2.12 shows the aspect of the Noachian terrain and the boundary between the Noachian-Hesperian terrain and Amazonian terrain.

Technical Box 4: Measuring Elevation on Mars
Determining the altitudes of the Martian surface is always of top importance for studies of geology and geomorphology, atmospheric circulation models, and planning of future missions, which require a good knowledge of the ground position. During the pre-Mariner age, it was difficult to determine altitudes, except for few hopeful guesses. Nix Olympica (Olympus Mons) had to be high for clouds to form at the top. It was only in the 1970s that a coarse DTM (digital terrain model) became available based on Viking images supplemented with Mariner data. The information resulted from a combination of different techniques, like image analysis, radar from Earth, occultation, and a study of the atmospheric pressure. Twenty years later, the MGS mission expanded our view of the planet's elevations by providing a complete altitude map with the infrared laser MOLA. Ten pulses were shot every second from the travelling spacecraft to the planet. Measuring their return time provided the precise distance at the reflection point. Because the spacecraft travelled around its orbit, and orbits change as the planet spins under the spacecraft, the whole planet was sampled in this way. The precision is excellent at the reflection points, but had to be extrapolated in between, which resulted in spotty precision. Resulting maps have a maximum resolution of 1/256 degree, which is sufficient for studies of the Martian surface that do not require a precision better than some hundred meters, but become blurred for studies requiring finer details. Figure 2.13 shows an example from Kasei Vallis. The optical (MOC) image is shown at the top, while the bottom figure shows the corresponding area contructed from MOLA data. Proper lighting is included in the digital model to simulate grazing sunlight coming from the top of the image. Note, however, that the magnified image appears indistinct (right).

A further step in the precision measurement of altitude was gained with the HiRISE camera on board the Mars Reconaissance Orbiter (MRO) and HRSC on board the Marx Express. These high-precision cameras can produce stereo pairs (i.e., photographs taken from slightly different positions) of selected details of Mars' surface. Our brain can estimate the distance of objects based on the different view that each of our eyes perceives. Similarly, stereo images elaborated with computer software can reveal the elevation on Mars' surface. Figure 2.14 shows an elevation image processed from a HiRISE stereo pair

(continued)

Technical Box 4 (continued)
from the same area in Kasei Vallis. Notice how altitudes are highly detailed, even though the area of the image (the rectangle in Fig. 2.14, right) is far below the level of MOLA resolution. Because a camera like HiRISE has a narrow view, only the altitudes of a few selected locations on Mars have been obtained, in contrast with MOLA data, which covered the entire planet.

SECOND MYSTERY: Was There Plate Tectonics on Mars?

Terrestrial tectonic plates, which are limited to the most superficial part of the Earth, the lithosphere, move horizontally, dragged by convection movements of the underlying mantle. The product of these slow movements is well visible in several geological processes, such as oceanic ridges, volcanism, and mountain ranges. Among the significant amount of evidence of plate tectonics, let us consider the Hawaiian Islands, not least of which because they offer a direct comparison with Mars: (more in Fig. 3.29). It is believed that a hot plume from the base of the mantle travels upward, driven by thermal convection. At the intersection with the crust, the

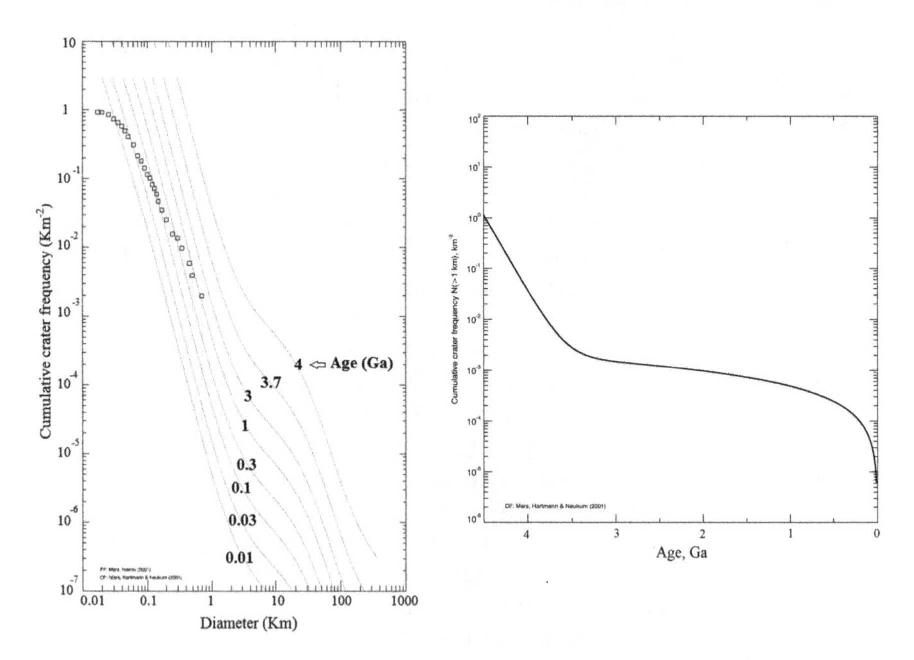

Fig. 2.11 The method of dating Martian terrains is based on crater-counting statistics. The abundance per unit of surface is then studied according to the diameter of the craters, and comparison with the calibration curves allows for surface dating. Here, the CraterStats program developed in Berlin is used. The example chosen is an area south of Acheron Fossae. Right: the number of craters per km^2 wider than 1 km as a function of the age of Martian surfaces. Image created with software CraterStats (G. Neukum and G. Michael)

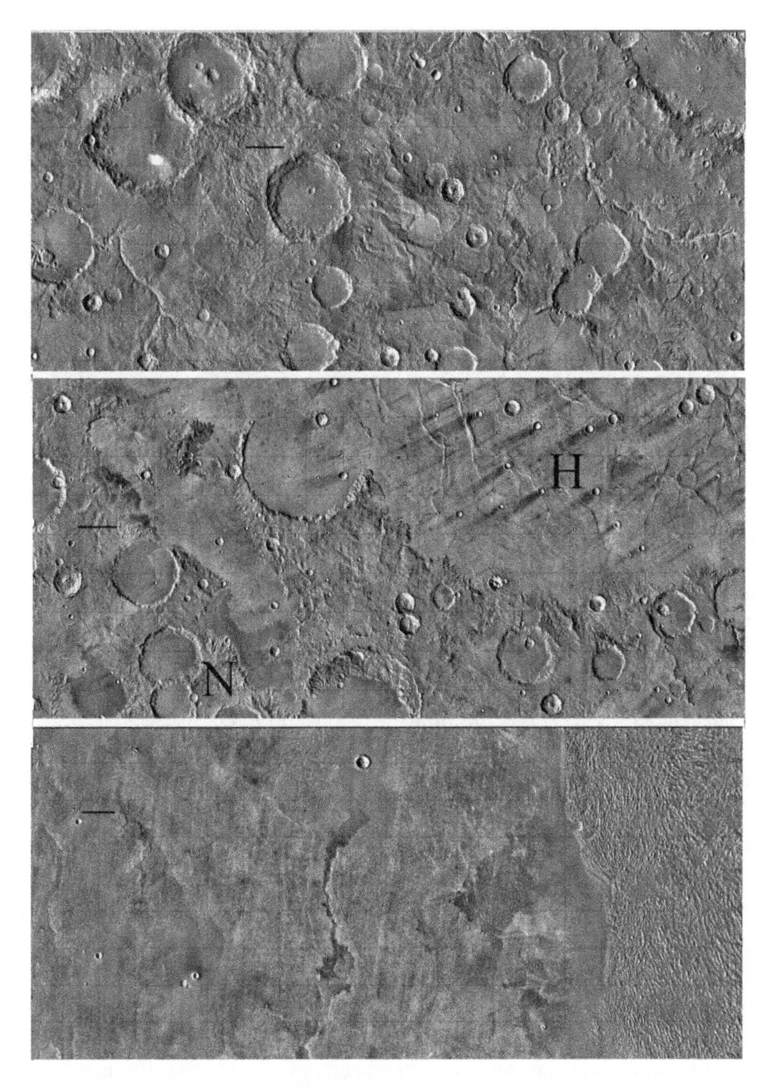

Fig. 2.12 Examples of terrains of the Noachian age (Terra Cimmeria, top), Noachian-to Hesperian transition (middle, Noachian indicated with "N," Hesperian with "H," transition between Syris Major Planum and Terra Sabaea), and the Amazonian (bottom, Amazonis Planitia). All figures approximately to the same scale (each bar is 30 km long). Note how the terrains of the Noachian age exhibit a far greater number of craters. A quick inspection of Martian imagery provides an expedited age guess indeed. Themis mosaic. THEMIS (Mars Odyssey), NASA, NASA/JPL-Caltech/Arizona State University

plume has produced the numerous volcanoes of the Hawaii chain. Since the plate moves northwest, the intersection between the plume and the ocean lithosphere changes over time. This explains the linear direction of the islands and the fact that the age of the islands decreases southward (the Hawaii chain goes further west to the Midway, and then points more northward in correspondence with the Emperor Islands, a consequence of the change of direction in the movement of the plate).

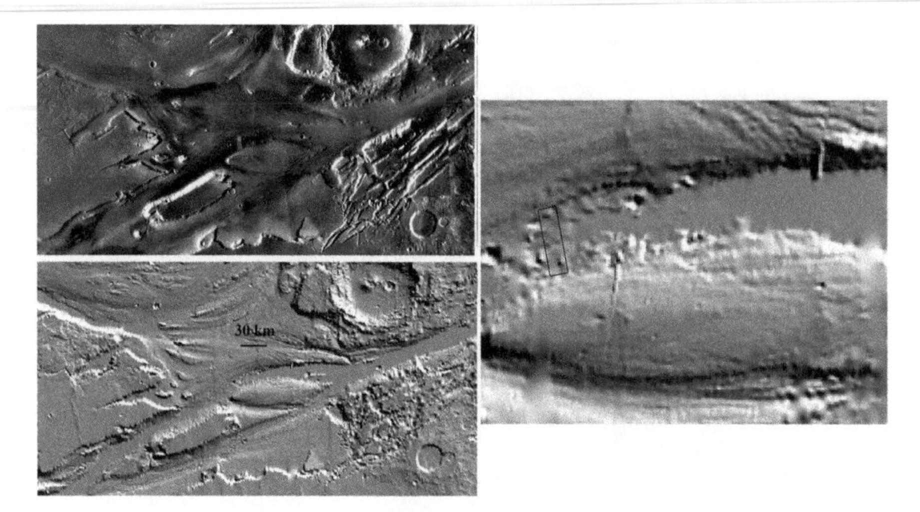

Fig. 2.13 A portion of Kasei Vallis, shown here as an example of altitude determination. Top: the MOC image. Bottom: the corresponding map derived from MOLA data. Right: magnified view of the "island" visible in the central part of the two previous images. The rectangle indicates the area shown next in Fig. 2.14. Top: MOC (Malin Space Science Systems, MGS), NASA, bottom and right: MOLA (MGS), NASA

Martian volcanoes are huge. Olympus Mons is 23 km tall, the maximum volcanic height recorded in the solar system. If there were plate movements on Mars, and if Olympus is the product of a plume as the Hawaiian volcanoes are, it would have formed a chain of volcanoes, and not a single volcanic edifice, which has consequently grown to extreme heights[6] (more on Olympus Mons in Chap. 3). The reason for the non-existence of present plate tectonics may be explained by a lack of thermal convection. Convection requires a number of conditions: high temperature difference between the base of the mantle and the lithosphere, low viscosity, and a thick sub-lithospheric mantle. The Earth benefits from all three of these aspects.

However, although evidence shows that plate movements were never of primary importance, some researchers have proposed that Mars temporarily developed some form of horizontal plate movement. It has been suggested[7] that the northern plains are the remains of an ancient oceanic area similar to the current oceanic crust on the Earth's surface. During the pre-Noachian, a seafloor spreading phenomenon comparable to present-day Earth may have occurred, with the ocean crust being subducted beneath the area occupied by the northern plains. Thus, the global dichotomy line would correspond to the active margins of terrestrial continents such as South America. The process was interrupted when cooling decreased the

[6]The volcanoes Arsia, Pavonis, and Ascreous Montes are lined up, and Alba Mons is also along roughly the same direction. It is believed, however, that this is so because of structural control (fracture lines).

[7]Sleep, N.H. 1994. Martian Plate Tectonics. J. Geophys. Res. 99, 5639–55.

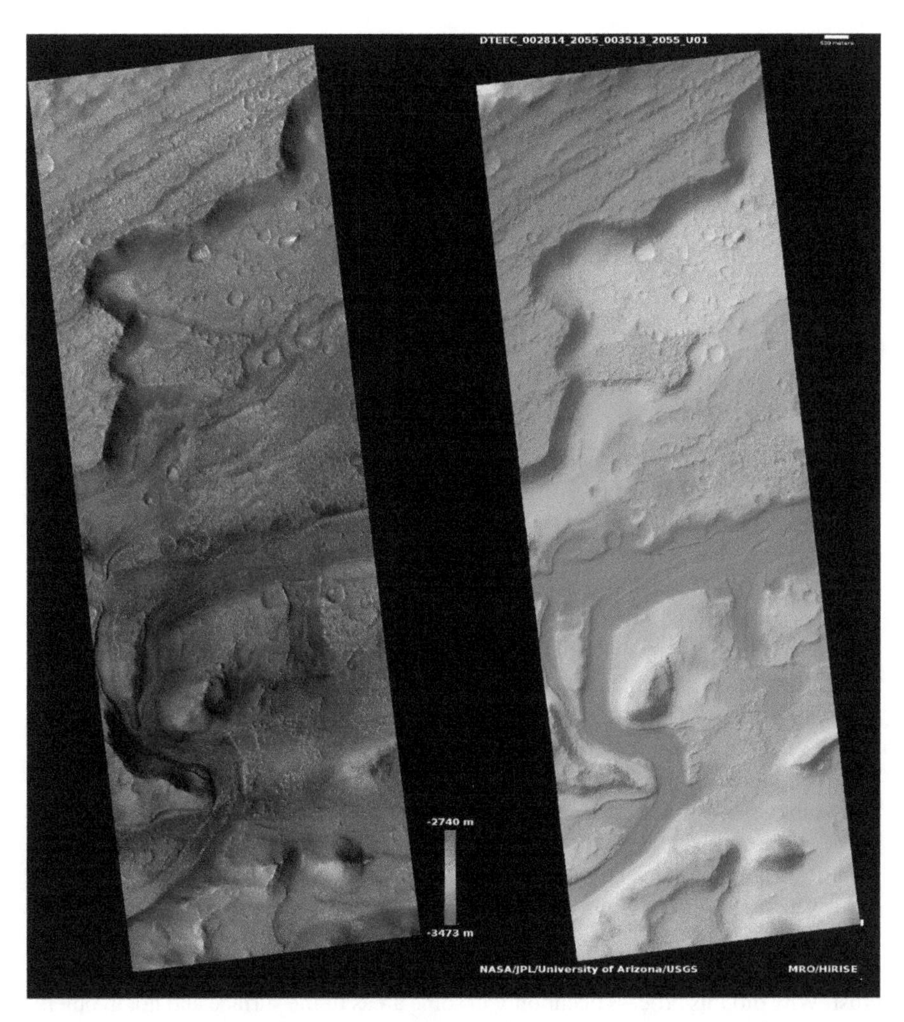

Fig. 2.14 Example of altitude reconstructed with HiRISE stereo image pairs, one of which is shown to the left. The artificial colors in the figure on the right indicate elevation from datum. The color bar is in the middle. The position of the image is shown as a rectangle on the right side of Fig. 4.13. Stereo pairs have also been produced by the HRSC camera aboard the European Mars Express mission. HRSC (Mars Express), ESA

energy available to form the convection cells in the Martian mantle, with the whole process initially being reduced and then no longer taking place.

From a completely different view, it has been hypothesized that plate tectonics would have been active in more recent, rather than ancient, times.[8] An indication

[8]Yin, A. (2012). An episodic slab-rollback model for the origin of the Tharsis rise on Mars: Implications for initiation of local plate subduction and final unification of a kinematically linked global plate-tectonic network on Earth. Lithosphere, 4(6), 553–593.

supporting this conclusion is a system of oblique faults oriented along a line crossing Valles Marineris by way of Coprates, Melas and Ius Chasmata. This horizontally-oriented line would offset the area to the north compared to the south by about 160 km. Since Valles Marineris is much newer than the formation of the global dichotomous line, presumably only tensions due to plate movements could explain the existence of passing faults. On the other hand, these structures are not as obvious as the terrestrial ones, and the plate tectonics of Mars (ancient or recent) still remains to be investigated.

2.3 Martian Dichotomy

Ever since Alfred Wegener's time, we have known that the Earth exhibits a marked diversity (dichotomy) between the continents and the oceans. The most obvious difference is altitudinal. The most common altitudes occur at sea level ("0" level), and at 5000 m below sea level (i.e., −5000 m). These two altitudes correspond to the average level of the continents and the bottom of the oceans (Fig. 2.15).

In addition to altitude, there is also a marked difference in composition between the oceanic and continental crusts. The continental crust, from about 20 to 50 km thick, consists of sedimentary, crystalline and igneous rocks. The continents are therefore a distillate of acidic rocks, rich in quartz and feldspar, contaminated with a veneer of sedimentary rock on the surface. The oceanic crust, on the other hand, is made of basaltic rocks resting over gabbroid rocks (gabbros are plutonic, i.e., non-eruptive rocks with the same composition as the basalt). Another difference between the continental and oceanic rocks on Earth is that the continental shields are pristine, while the oceanic crust is continually created in the ocean's ridges and destroyed in the subduction zones, in a continuous cycle explained by the plate tectonics. The terrestrial dichotomy between continental and oceanic rocks therefore dates back to ancient times, when gravitational differentiation formed the continental crust. Why did this process occur only in some areas of the surface and not in others?

Although, as we have discussed, Mars currently shows no evidence of plate tectonics, the distribution of altitudes shows a marked dichotomy that recalls that of the Earth (Fig. 2.15[9]). Similarly to the Earth, the two peaks at the highest and lowest altitudes correspond to the southern plateaus and the northern lowland, respectively. In other words, simplifying a bit, it appears as if the south of Mars is "continental" and northern Mars is "oceanic." There is, however, a big difference from the Earth, where continents are not confined to one hemisphere, but rather are scattered around the globe. Figure 2.16 shows a south-north section through one

[9]Because Mars is currently devoid of oceans, there is no "zero" level for altitude. Initially, a zero level was defined based on the atmospheric pressure. New altitude measurements with the MOLA laser on board the Mars Reconaissance Orbiter mission have led to a redefinition of altitude "0" as a distance from the center of the planet equal to the average radius. Pay attention to this double definition of altitudes, as they are both in the literature.

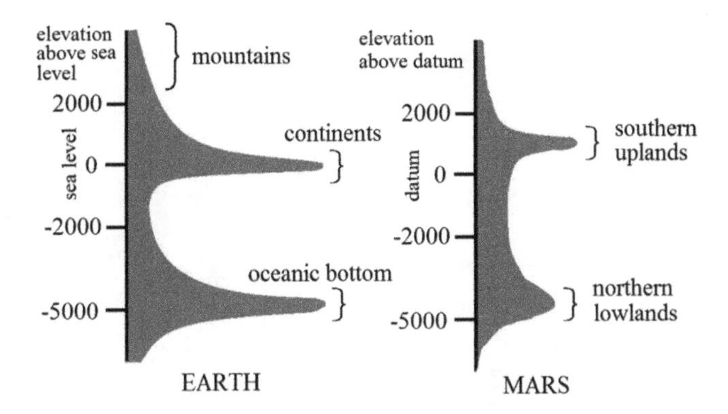

Fig. 2.15 Hypsographic distributions of Earth and Mars. FVDB

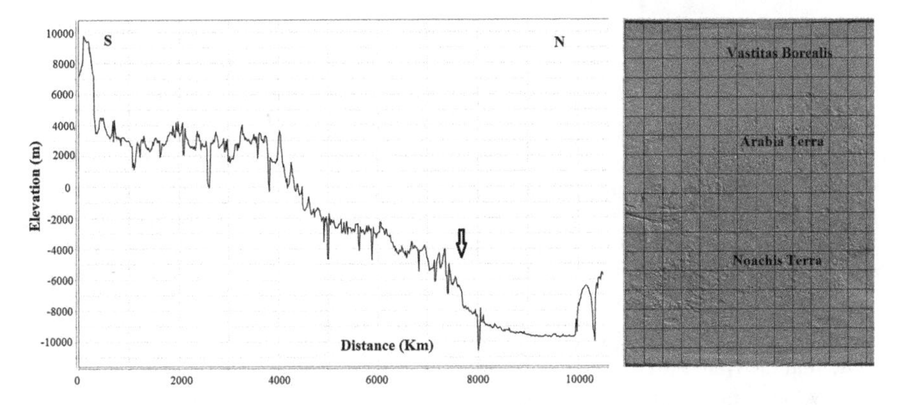

Fig. 2.16 MOLA elevation profile through the zero point of longitude intersecting Noachis Terra, Terra Sabeaea, Arabia Terra, and Vastitas Borealis. The position of the dichotomy is shown with an arrow. MOLA (MGS), NASA

chosen Martian meridian, with the dichotomy point indicated by an arrow. The dichotomy appears as a small "kink" in the profile, showing that the transition is not abrupt.

The second difference between northern and southern Mars is that the southern highlands show much higher crater density than the northern plains. The MOLA altimeter laser on the MGS spacecraft was the first probe to show the immensity of Utopia Planitia's depression on the northern plains, where the lander from Viking 2 had disembarked (Fig. 2.1). In the northern plains, numerous large "ghost" craters have been described. These are termed quasi-circular depressions (QCD), a class similar to the impact basins, but much less visible because they are very shallow. In tracing these QCD, gravimetry has also been used by measuring the variation of the MGS spacecraft velocities in its orbit around Mars. In fact, hidden masses and variations in density and mass distribution within Mars would show up as orbital

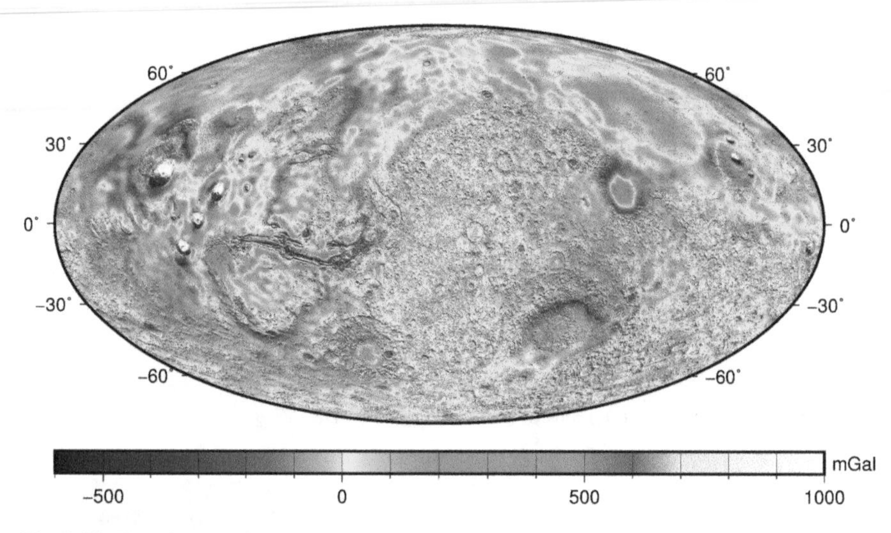

Fig. 2.17 Free air anomaly map reconstructed by data processing of the orbits of the spacecrafts MGS, Mars Odyssey and MRO. The map takes into account the necessary global corrections (the increase of Mars' radius with latitude should be subtracted, as it gives a gravity difference of 0.04 m/s^2 or 4000 mGal, between the equator and the poles), and for the local differences in height due to topography. Courtesy of the Goddard Space Flight Center. For this and the next map, see ref. (Genova, A., Goossens, S., Lemoine, F. G., Mazarico, E., Neumann, G. A., Smith, D. E., & Zuber, M. T. (2016). Seasonal and static gravity field of Mars from MGS, Mars Odyssey and MRO radio science. *Icarus, 272*, 228–245). Various probes, NASA's Scientific Visualization Studio (A. Genova, E.M. Mazarico, E. Wright, D. Gallagher)

anomalies of the Mars Odyssey and MGS orbits. Figure 2.17 shows such a map.[10] Note that the final gravity on the top of Olympus Mons is smaller, since the volcano is very high and its surface is more than 20 km further from the center of the planet than the surrounding areas. Positive anomalies (red color) are associated with large impact basins, such as Utopia and Isidis Planitiae in the north. We also notice the traces of large QCDs as colored semicircles. In addition to the largest QCDs, there are also smaller depressions several kilometers or tens of kilometers wide, but only a few meters deep. The contrast between such a very tiny depression and the surrounding terrain is undiscernible in optical images, but well visible with MOLA altimeter data. This is so because MOLA allows for discriminating differences in elevation of only a few meters, provided that they are extended in space. In this way, stealth QCDs have been found.[11] However, dating the northern lowlands on safer ground implies counting only large depressions. Thus, the integral density of QCDs wider than 200 km (i.e., the number of QCDs larger than 200 km divided by the area

[10]To be precise, it shows the free air anomaly, that is, the correct "true" local gravity, which requires some corrections. Because the polar radius is smaller than the equatorial radius, Mars' gravity at the surface increases globally towards the poles. In addition, gravity is also influenced by the rotation due to the centrifugal effect and by local topography.

[11]D.L. Buczkowski, Comparing quasi-circular depression (QCD) locations to northern lowland materials on Mars.

of the northern lowlands) is about 0.0000025 km^{-2}, or in power notation, 2.5 10^{-6} km^{-2}. This would correspond to an age of the northern lowlands of about 4.1–4.2 Ga (See Technical Box 3).

These measurements of the gravity at the surface of Mars have pinpointed yet another significant difference between the planets' southern and northern hemispheres. Based on the local gravity, and using a constant density model for the crust, it has been found that the crust is thicker in the southern part of Mars, where it would reach a thickness of about 60 km, twice the thickness in the northern basins (Fig. 2.18). It is likely that the northern craters have been heavily buried by a thick blanket of sediment.

Finally, we look at the smoothness of the northern lowlands compared to the rough southern part of Mars. This is evident looking at Fig. 2.19, which shows the roughness of the whole planet. Notice how the highlands in the south appear whitish compared to the highlands in the north (Fig. 2.19).

THIRD MYSTERY: What Is the Origin of Global Dichotomy?

Theories to explain the origin of Martian dichotomy fall along different conceptual lines. It has been hypothesized that the crust was originally thinner on one of the two hemispheres.[12] Due to isostasy, the floating lithosphere in the thin hemisphere (which subsequently became the northern lowlands) readjusted to the lower altitude. At this point, the northern lowlands were blanketed with sediment. Because of the enormous amount of sediment accumulated, the aspect and smoothness of the lowlands was changed, and in a planet devoid of rock recycling, such changes have remained in place to this day. We have already mentioned the possibility that plate tectonics may have occurred on early Mars, and that according to some theories, the dichotomy line would correspond to the active margins on Earth, where oceanic lithosphere is subducted beneath continental lithosphere. When plate tectonics ceased in the Noachian, the sub-circular margin was frozen into the current shape. However, no geophysical or morphological evidence in favor of subduction on Mars exists.

According to a second group of theories,[13] Mars was initially covered with a uniform crust. It was only later that some process starting from the core of the planet removed the crust in correspondence with the northern plains. Perhaps this extraordinary erosion event was controlled by the convection currents in the Martian mantle[14] or, alternatively, the crust was overturned and swallowed in the mantle.[15] To clarify this latter process, let us recall that a glacial lake begins to solidify from

---

[12]Mutch, T. A., Arvidson, R. E., Head III, J. W., Jones, K. L., & Saunders, R. S. (1976). The geology of Mars. *Princeton, NJ, Princeton University Press, 1976. 409 p.*

[13]McGill, G.E., and Squyres, S.W. 1991. Origin of the Martian Crustal Dichotomy: Evaluating Hypotheses. Icarus 93, 386–393.

[14]Wise, D. U., M. P. Golombeck, and G. E. McGill 1979. Tectonic evolution of Mars. J. Geophys. Res. 84, 7934–7939.

[15]Elkins-Tanton, L. T., Hess, P. C., & Parmentier, E. M. (2005). Possible formation of ancient crust on Mars through magma ocean processes. *Journal of Geophysical Research: Planets*, *110*(E12).

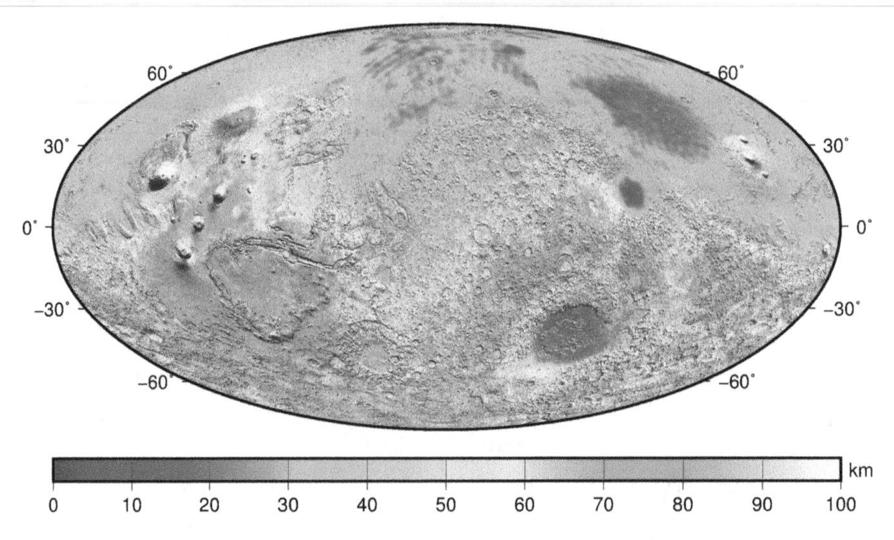

Fig. 2.18 Crustal thickness (superimposed onto MOLA shadowed relief) obtained by gravity free air anomaly and crustal density of 3500 kg/m^3. Various probes, NASA's Scientific Visualization Studio (A. Genova, E.M. Mazarico, E. Wright, D. Gallagher)

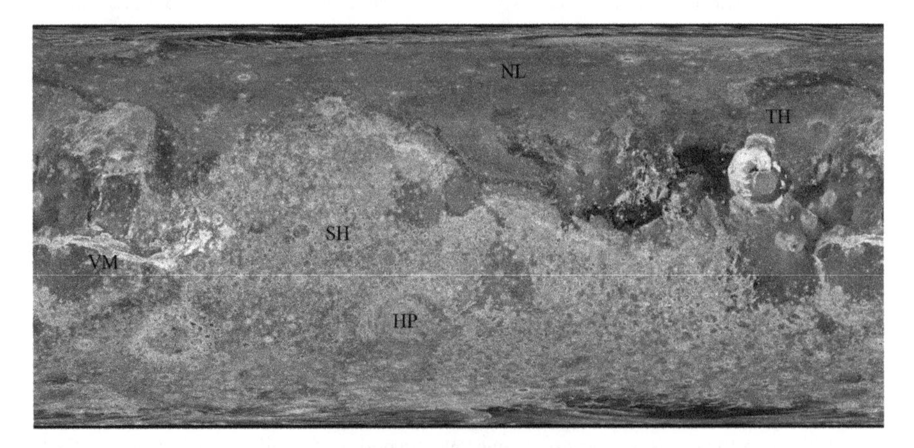

Fig. 2.19 Smoothness of the Martian terrains calculated with 2.4 km of baseline MOLA data. The Northern lowlands (NL) appear much smoother than the Southern Highlands (SH). Also indicated are the Tharsis volcanic region (TH; note the very rough aureole, appearing as a whitish halo), Valles Marineris (VM) and Hellas Planitia (HP). MOLA (MGS), NASA, JMARS platform

the surface. Since ice is less dense than water, it forms a superficial crust that does not sink, but rather continues to increase in volume at the expense of water. Molten magma can also form lakes, such as the ones in the crater of the Kilauea volcano on Hawaii. However, when a solid crust forms on top of these magma lakes, it sinks, as the solid lava is denser than the underlying molten magma. We know that the solid planetary crust is a magma distillate. Following this line of reasoning, it has been hypothesized that the magma ocean originally present on Mars crystallized in such a way that cumulates (aggregates of heavy minerals such as olivine) could accumulate above the upper mantle, forming a layer heavier than the mantle itself. At that point, an overturn resulted in those areas predestined to become the northern plains. However, it is unclear why there should be an overturn in only one hemisphere.

Yet another hypothesis proposes the dichotomy as being due to processes internal to Mars. In analogy with the Earth, it is believed that there was a time when the mantle of Mars was in a state of vigorous thermal convection. Numerical simulations of thermal convection in planetary mantles show that the convection patterns can be very varied depending on the temperature gradient, the thickness of the mantle, and the viscosity. A limit case of convection occurs when there is a strong variation of viscosity through the Martian mantle, because in this case, a single convection cell is formed. Convection currents rise in the upwelling areas corresponding to one whole hemisphere, and downwelling makes the mantle fall into the other.[16] This process would thus generate a pristine difference between hemispheres. A similar hypothesis is that a strong density contrast may form a superplume, rather than a convection cell (a superplume is a hot rising region of the mantle shaped like a mushroom) in only one hemisphere. If these internal models succeed in reproducing some partial difference between the two hemispheres, the details are unclear.

We do not know any model, even a partial one, capable of totally explaining global dichotomy as a result of the inner activity of the planet. The alternative is this: was it a process that came from above, and not from below? Could a catastrophic impact be the solution to the dilemma? We have seen how enormous impacts have affected the history of all planets. Also, on the Earth, such celestial bodies have fallen continuously, but our planet has been very quick to heal its surface scars. The Moon was created by the collision of a huge planetary body with Earth, which scoured its mantle. Why couldn't a similar dramatic event have occurred on Mars? History seems simple in its dramatic nature. At first, the Martian crust was formed, uniform throughout its surface. The impact with a planet-size body (between 600 km and 1000 km in diameter) scraped away one-third of the Martian crust for a total volume of one billion cubic kilometers, exposing the underlying materials. Since there is no rock recycling on Mars, a scar of this magnitude would remain for over four billion years, bearing witness to that primordial drama, blanketed with sedimentary infilling the northern plains (Fig. 2.20).

A first simple objection concerns the irregular periphery of the northern plains. We have seen that a meteorite impact tends to be circular, while the northern plains have an elliptical shape (to get an idea of the dichotomy line, we have to imagine how the line might have looked prior to more recent morphology, especially due to the Tharsis volcanic area). According to some authors, one indication of a giant impact is that the northern lowlands appear to be an elliptical basin,[17] the largest axis

[16] Zhong, S., & Zuber, M. T. (2001). Degree-1 mantle convection and the crustal dichotomy on Mars. *Earth and Planetary Science Letters*, *189*(1–2), 75–84.

[17] Andrews-Hanna, J. C., Zuber, M. T., & Banerdt, W. B. (2008). The Borealis basin and the origin of the martian crustal dichotomy. *Nature*, *453*(7199), 1212–1215.

Fig. 2.20 According to the giant impact hypothesis, the northern lowlands were created by one impact of a body between 600 and 1000 km in diameter. An artist's conception of a giant impact between two planetary bodies. Image 17354137 FOTOLIA/sdecoret

of the ellipse being approximately 25% longer than the lower axis. Thus, perhaps the direction of impact was skewed. It seems, in fact, that the rule according to which an impact crater is usually spherical is no longer valid for such huge craters. Hellas Planitia itself has the shape not of a circle, but of an ellipse whose largest diameter (directed east-west) is about one third larger than the smaller diameter. It has been speculated that perhaps the curvature of the planet may play a role in creating elliptical impact basins when the diameter of the asteroid becomes comparable to that of the main planet.

During impact, the tangential velocity component of the impactor must have been communicated to the planet, which modified its rotation speed around its axis. Suppose, for simplicity, that before the impact with Mars, the asteroid had a speed of rotation of zero. It is easy to calculate, based on the equations of conservation of the angular momentum, that the impact must have placed the planet rotating with a period of 35–100 h if the impactor had a radius of 500 km. Today, Mars rotates at an angular velocity similar to that of the Earth. So, if the impact really took place, it would have altered the angular velocity and rotational kinematics of Mars, but it is very difficult to say how. Another hypothesis is that the giant impact did not occur in the (present-day) northern hemisphere, but rather in the southern hemisphere, and thus the product of the impact would not be the missing crust, but the crust itself in the south. According to this hypothesis, the impact created (or re-created) a magma ocean from which the southern crust resulted after cooling and solidification.[18]

The other considered problem, that the perimeter of the impact basins is normally a continuous line with little sinuosity while the dichotomous line appears irregular, emphasizes the difficulty of the notion that a single, big impact could have created such a bizarre form. It has been suggested that the asteroids that created the northern lowlands were many, and not just one, and did not necessarily impact Mars at the same time. This theory could explain the elliptical shape, but also opens up other difficult problems. The first problem is statistical. The northern plains are about one-third of the planet. Let us suppose that a first big asteroid falls, creating a crater, say, a 1000 km in diameter. The second asteroid must have fallen from the same part of the planet to create a final depression on that side alone. The probability that this would happen is (1/3) (1/3) = 1/9. In order to make a depression the size of the northern plains, a number of asteroids would be required at least equal to the total area of the northern plains divided by the area created by one impact, but probably many more, since the impacts would partly overlap one another. At least 5–7 impacts on the order of those that produced Utopia Planitia, or many more for smaller asteroids, would be needed. The probabilities of all of this are very low.[19]

Another objection seems even stronger. As many researchers have acknowledged, although the northern plains have a smooth appearance, they actually host large depressions, called quasi-circular depressions (QCDs). Such depressions are hardly visible due to the presence of thick rocks. Utopia Planitia and Chryse Planitia are the most obvious, but they not the only ones. Even the enormous Utopia Planitia, of a diameter comparable to that of the basin of Hellas Planitia, seems to be an accident far more superficial than could possibly have contributed to the formation of a global depression. And furthermore: where have the ejecta of such a colossal impact or impacts been cast that they could have ripped off a six-kilometer-thick cover? If these ejecta were redistributed around the northern plain within a ring of 500 km, they should have a thickness of at least 70 km. A mountain range of this altitude around the northern plains would certainly be noticeable. And even if the material had been redistributed uniformly throughout the planet, it would have raised its average by a couple of kilometers. For sure, the problem of the Martian dichotomy, once resolved, will reveal a fundamental aspect of early Martian history.[20]

---

[18]Leone, G., Tackley, P. J., Gerya, T. V., May, D. A., & Zhu, G. (2014). Three-dimensional simulations of the southern polar giant impact hypothesis for the origin of the Martian dichotomy. *Geophysical Research Letters*, *41*(24), 8736–8743.

[19]The likelihood that all of these asteroids have fallen on the same part is equal to $(1/3)^n$ or less than one in two thousand (if they have fallen all over n = 7 very large asteroids) up to one in sixty thousand (if they fall n = 10 slightly smaller asteroids).

[20]For a detailed review, see McGill, G.E., and Squyres, S.W. 1991. Origin of the Martian Crustal Dichotomy: Evaluating Hypotheses. Icarus 93, 386–393.

Addenda

- Martian morphologies are identified by two designations: the first indicates the type of morphology; the second is taken from Greek–Roman mythology or other sources.
- The highest point on Mars is reached at 21,287 m at the top of the Olympus Mons volcano. The lowest elevation is recorded inside Hellas Planitia, at −8180 m.
- The Pathfinder mission of 1997, which, together with the Mars Global Surveyor, brought the focus of technology back onto Mars after twenty years of senescence, cost less than the $ 175 million used to make the blockbuster "Waterworld," which was released in the same period.
- From a biological point of view, the role of the Earth's magnetic field is essential to protecting life from the lethal flow of solar cosmic rays, diverting them along trajectories far from the Earth's surface. Without this invisible field, life as we know it could not have developed.
- The meteorite from Mars, ALH84001, has a crystallization age of 4091 Ma and is still magnetized, which implies that the Martian magnetic dynamo was still active at that time.
- Mars and Earth have two similar orbital characteristics: the axial tilt and the rotation period. Because the Martian tilt angle varies wildly and quickly, and since in the past, the Earth rotated faster, the present similarity is probably accidental.

Chapter 3
The Surface of Mars

On Earth, the formation of mountain ranges as the consequence of geotectonic cycles, the deposition of sediments on continents and oceans, the eruption of volcanoes, and catastrophic phenomena such as earthquakes and landslides has continued incessantly. By accumulating new rocks and creating new morphologies, the geological forces have changed the layout of the preexisting ones. Meanwhile, erosion due to rivers, glaciers, thermal effects, wind, and also biological activity have quickly covered or deleted the traces of earlier geological structures in a never-ending cycle. As a result, the landscape on Earth is often geologically recent. But on Mars, both the superficial expression of new geological phenomena and the erosive phenomena have been much slower. The lack of plate movements and the absence of life, at least in terms of terrestrial characteristics, have aided in the conservation of traces of geological activity after extremely long periods. Just as a family picture shows people of more generations, images of Mars often display morphologies that are billions of years old, along with much more recent features. On the following pages, we will see many astonishing images of the Martian surface. We will investigate the nature of huge meteorite impact craters, appreciate what they can reveal about the properties of the Martian terrain, examine the highest volcanoes in the solar system, and be surprised by the volume and length travelled by landslides as large as the whole of France. We will see structures that, surprisingly, often appear very close to each other, and are much neater and tidier than those on Earth.

F. V. De Blasio, *Mysteries of Mars*, Springer Praxis Books,
https://doi.org/10.1007/978-3-319-74784-2_3

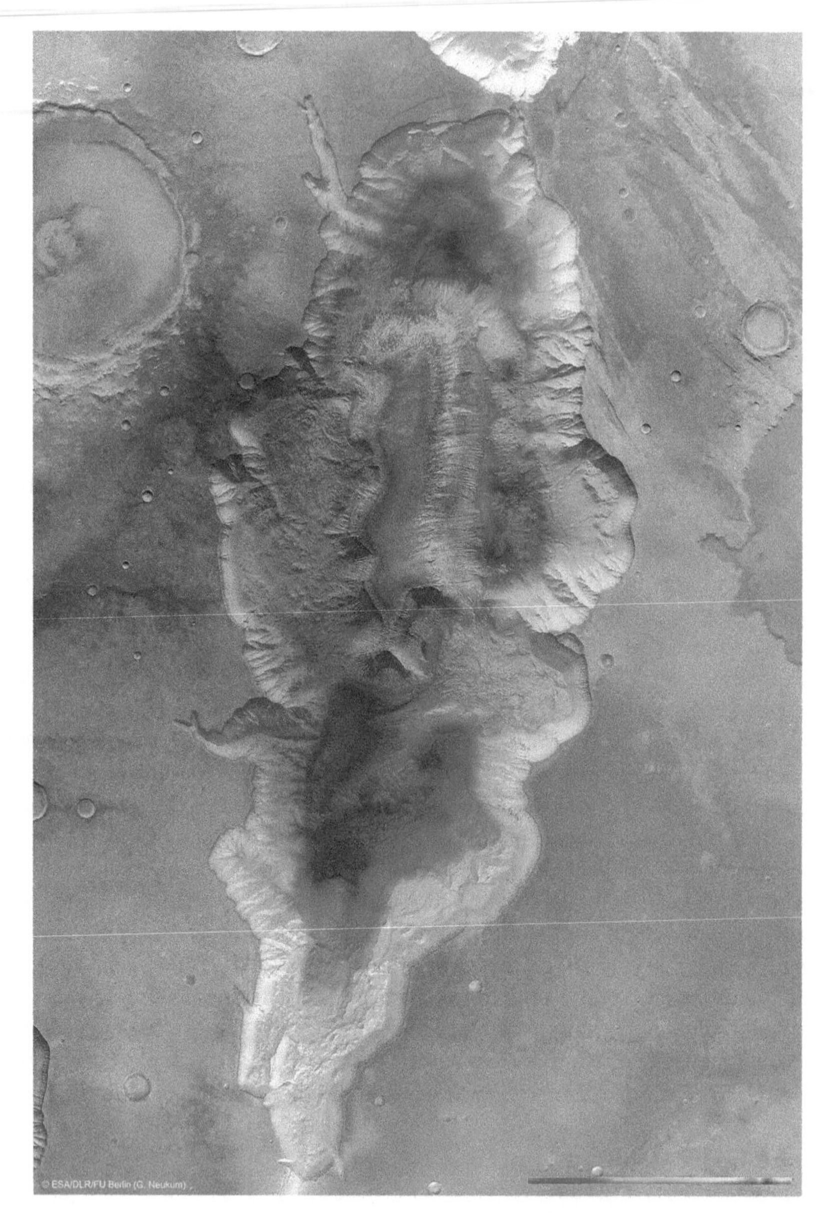

Hebes Chasma is a closed chasm on Mars showing headwall degradation, landslides, sapping channels, and glacial features. In this amazing topography image, each color corresponds to a certain elevation (blue and purple the lowest, white and red the highest). It was obtained by processing eight different HRSC images (High Resolution Stereo Camera) on ESA's Mars Express. Courtesy of ESA/DLR/FU Berlin (G. Neukum). ID 297913, north is to the right. HRSC (Mars Express) ESA/DLR/FU Berlin (G. Neukum)

3.1 Impact Craters

Figure 3.1 shows Manicouagan's giant crater in Qebec, created by the fall of a 10 km-wide asteroid about 212 million years ago. It was the mid-late Triassic, and the first small forms of dinosaurs were making their appearance on an Earth warmer than today. Over two hundred million years ago may appear to be a long time, but it is only one twentieth of Earth's age. The large circle about 70 km wide in the image, made visible by the lake water, is just the inner circle of a double ring impact basin. The outer ring, 100 km wide, can only be discerned by field survey.

In a geologically brief time, the crater was eroded very deeply by water and ice. Thus, the example of this crater shows that on Earth, the landscape changes rapidly. The reasons are fourfold. First, Earth is a very geologically active planet, and volcanoes, earthquakes, mountain chain formation, and related phenomena tend to erase the traces of impact craters. A second factor is erosion: on Earth, rivers, glaciers, thermal effects and wind are very efficient, although impressive traces of the same erosive agents can also be identified on Mars. The third reason has to do with life. Forests, the activity of animals, plants and fungi, combined with the presence of water to form a layer of soil, contribute to the erosion and the continuous recycling of surface geological materials. Finally, because horizontal plate movements continually recycle the ocean crust, the bottom of the oceans is especially young from a geological viewpoint.

On Mars, erosion and deposition has been much less accentuated. Consequently, many geological forms can coexist in the same area, often showing little change after billions of years. Let us take a look at meteorite impact craters on Mars, one of the most noticeable forms on the Martian surface.

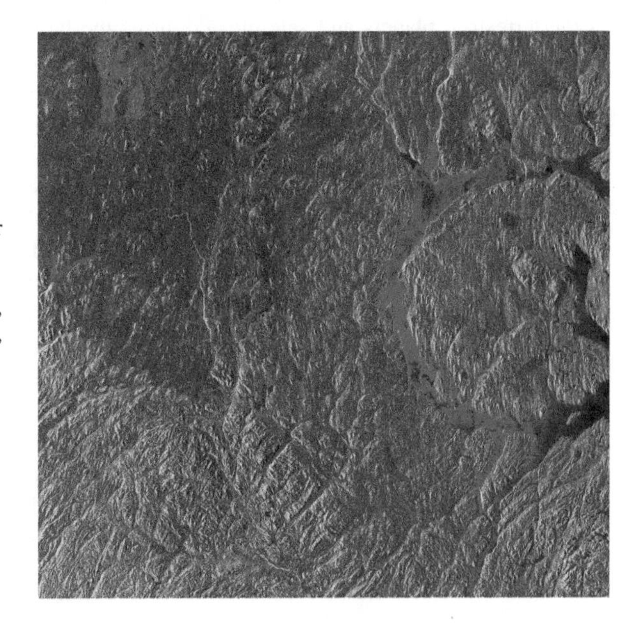

Fig. 3.1 False-color image of the Manicouagan astrobleme (impact crater) taken by the Sentinel-1A satellite of ESA. Vegetal and animal life have been very efficient in changing the morphology of the crater (212 million years in age, with an uncertainty of plus or minus four million years), together with glacial, fluvial, and other forms of erosion. Well visible is the inner, 70 km-diameter ring of the astrobleme. Image taken by Sentinel-1A satellite, artificial color composite. Sentinel-1A satellite, courtesy ESA

Fig. 3.2 The hidden side of the Moon. NASA. Courtesy NASA

Impact Craters as a Planetary Phenomenon

According to the planetologist Eugene Shoemaker, the impact of solid bodies from space is the most fundamental of all processes that have taken place on the terrestrial planets (Mercury, Venus, Mars and, obviously, the Earth). A simple view of a map of the terrestrial planets and satellites confirms the large number of craters on the surface of the planets and the importance of impact cratering. Also, the assorted moons and planetoids are completely covered by craters. Our Moon is the most classic example. There are huge impact basins, concentric craters, craters with radial ejecta, but also small craters of just a few tens of meters in diameter. Figure 3.2 shows the hidden side of the Moon. The craters are so dense that they overlap each other, a situation called saturation. It is an indication of the huge number of impacts that have tormented the Moon during its nearly five and a half billion years of life. Impact cratering has also affected the surface of Mercury, while it had a somewhat reduced role on Venus, where the thick atmosphere provided a protective shield, disintegrating the smallest impactors before they could make it to the ground.

As mentioned, some regions of Mars are full of meteorite craters, while others are rather lacking. The Noachian terrains are rich in craters of various sizes, but crater density decreases on the Hesperian and especially the Amazonian terrains. However, so far, we have considered craters as simple holes without structure. Here, we will consider the mechanics of impact cratering in more detail, investigate the large variety of Martian craters, see what they can reveal about the properties of the Martian terrain, and make a comparison with similar terrestrial structures.

Impact Craters on Mars

Figure 3.3 shows some "classic" examples of Martian craters, ranging from large ones to smaller ones. The depth of a crater on Mars increases with its diameter, but not in a proportional manner. A crater 100 m in diameter has an average depth of 20 m, one 10 km, a depth of 920 m. As the crater size increases, the depth tends to saturate: a crater 100 km in diameter has a typical depth of 3.7 km. And the shape also changes with size. While smaller craters are shaped like bowls, larger craters have more complex morphologies, terraced walls, and a central peak. Also, the bottom of the crater becomes flatter. There are also the ejecta, that is, the material ejected from the bottom and projected onto the sides of the impact. On the Moon, the ejecta have formed a deposit that becomes thinner away from the crater. On Mars, however, ejecta often take peculiar forms.

In the polar regions, and particularly in the Northern Hemisphere, so-called *pedestal craters* are relatively common, the ejecta pedestals of which rise from the planet's surface. Typically, such ejecta typology is frequent between 45 and 60° of latitude. One kind of surprising, exclusively Martian class of craters is the *layered ejecta craters*. As the name suggests, the ejecta form a blanket of rocks, reaching a

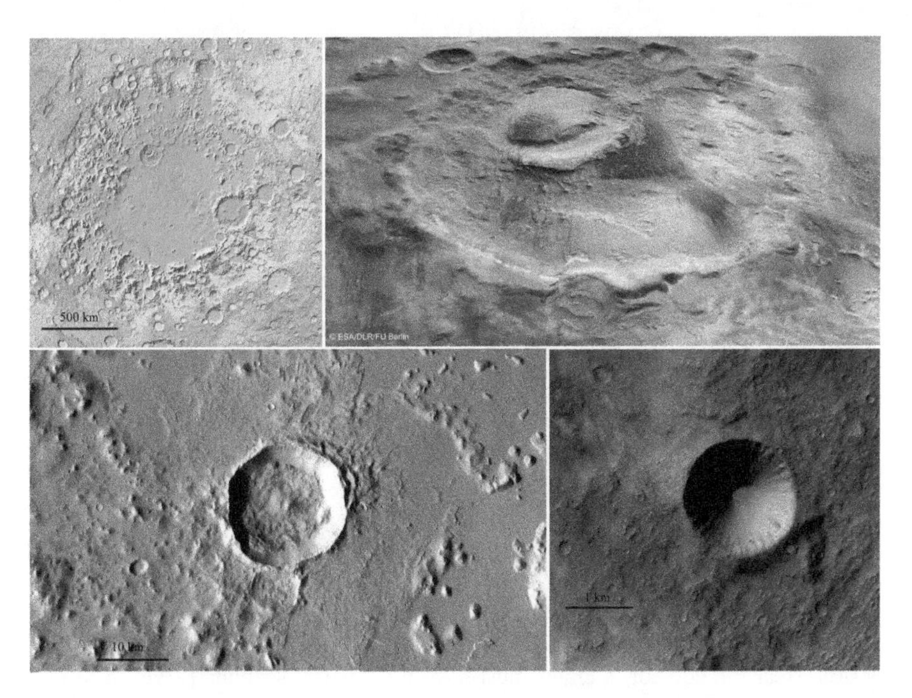

Fig. 3.3 Scaling down craters on Mars. Top west: Argyre Planitia, 1500 km in diameter. MOLA colorized elevation map. Top right: The Hooke crater inside Argyre, 150 km across. Courtesy of the ESA Mars Express, image HRSC 322700. Bottom, left: a crater 15 km across located at 13°16′, 165°17′ E. Image CTX composite on Google Mars platform. Bottom, right: crater 1.5 km across. CTX image. Coordinate −10°51′, 83°05′ E. MOLA (MGS), NASA, HRSC (Mars Express, ESA), CTX (MRO), NASA, on Google Mars platform, CTX (MRO), NASA

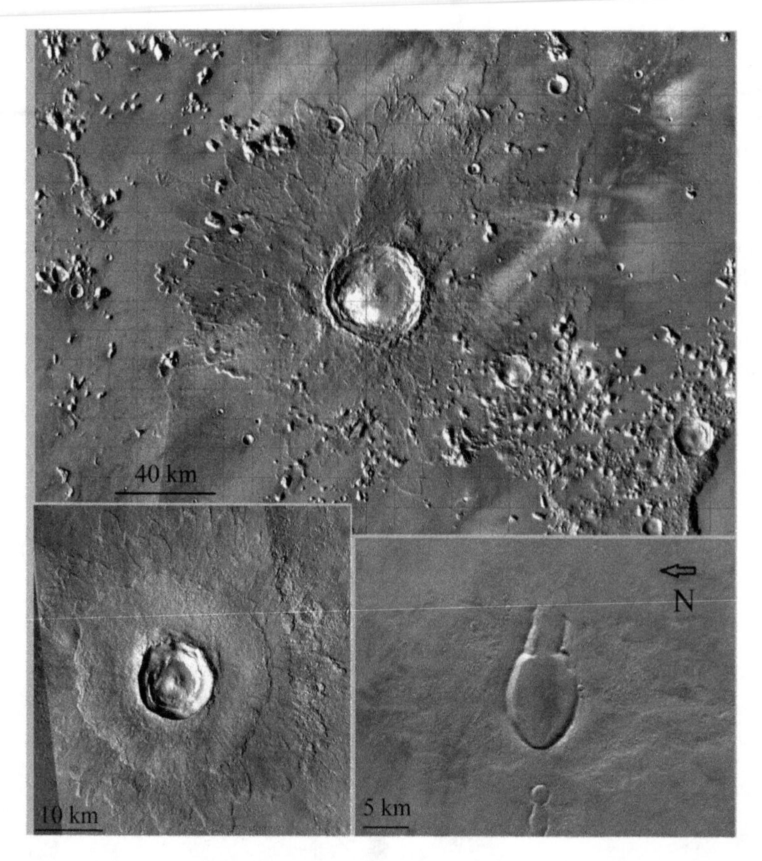

Fig. 3.4 Top: multiple-layered ejecta: the Kokta crater. Image THEMIS (Mars Odyssey); bottom left: double-layered ejecta crater: the Steinheim crater. Image CTX (MRO) on Google Mars platform. Bottom right: double-layered ejecta, oblong crater due to skewed impact with its layered, specular symmetric ejecta. CTX (MRO). THEMIS (Mars Odyssey), NASA, CTX (MRO) on Google Mars platform, CTX (MRO), NASA

length of tens of kilometers. The ejecta appear like a fluid sheet that has suddenly frozen. Different types of layered ejecta crater can be distinguished. Some craters exhibit a single layer of ejecta (SLE). Others are endowed with double-layered ejecta (DLE), and there are even those with multiple layers (MLE). Figure 3.4 shows a sample of these craters. One interesting thing is that, although these layered ejecta craters are distributed throughout the planet, ejecta craters are observed only around craters of sufficiently large diameter. Craters that are too small are not associated with ejecta layers. The critical size depends on latitude, especially for DLEs and MLEs. Near the equator, a crater must have a diameter of about 12 km to form a DLE. At 60° to 70° latitude, however, the minimum diameter drops to 5–6 km. For MLE, the critical diameters are about 20–25 km at the equator, and 12–15 km at the high latitudes. We shall return to this point in Chap. 4, where we will see how these details reveal surprising features of the Martian soil.

Technical Box 5: Meteorite Impact and Crater Formation

For a long time, it was discussed whether the Lunar and Martian craters were meteoritic or volcanic. Although the volcanic faction was never overwhelming, some of their arguments seemed valid. It was not only because some terrestrial volcanic craters resemble lunar craters. More importantly, if craters were the result of meteorite falls, many of them should have an oblong shape, as a result of a skewed impact. For what reason, then, are almost all of the craters nearly circular? In fact, there are elliptical craters on Mars. Some of them are very deformed, with the largest axis being two or three times larger than the shorter one. Figure 3.5 shows perhaps the most picturesque example, the big Orcus Patera in the Elysium region. Based on statistics, there seems to be a continuous distribution of the aspect ratio between the two axis lengths of an impact crater (Fig. 3.6). However, craters like Orcus are rare. We know today that unless the angle of impact is small (below 5°), the craters tend to be circular regardless of the trajectory of the falling meteorite. Let us then see what happens during a very rapid impact on a planetary surface.

Knowledge of impact cratering mechanisms has been greatly advanced in recent years, based on studies of impact craters on Earth and other planets, explosion experiments, and shock wave measurements with high-speed bullets. When a meteorite strikes a planetary surface, there is initially a compression phase, during which the kinetic energy of the rock is transferred in a very short time to the planetary surface. Let us consider the case of an iron meteorite with a diameter of 2 km. The meteorite mass, assumed to be spherical, is equal to 3.36×10^{13} Kg using a density of 8000 kg/m^3. At an impact velocity on the order of 10 km per second, the kinetic energy becomes 1.6×10^{21} joule (1000 joules correspond to about a quarter of a kilo-calorie). Since the explosion of one ton of TNT gives off 4.2×10^{12} joules, the meteorite's energy equals four billion tons of TNT, or four million kilotons, equivalent to about 250 Hiroshima bombs.

During impact, pressure on the meteorite front rises enormously, reaching levels of hundreds of gigapascals, or millions of times the atmospheric pressure on Earth. A shock wave is generated across the planet, while the meteorite-related material and the meteorite itself first crumble, and then vaporize. Splashes of fluidized material are thrown from the impact zone at very high speed (a phenomenon known as jetting). Deeper, the transformation of planetary material is not so radical, but is still sufficient to create new mineral phases. This first phase of impact, called the compression stage, takes the time required for the meteorite to move about ten times its diameter. For a meteorite 1 km in diameter, this time is about a second.

In a further step, the crater bowl is created. The shock-induced impact wave propagates within the planet at speeds of 6–10 km/s, faster than the speed of sound in the planetary rocks. In the nearest parts, a few 100 m deep, the rock is metamorphosed, transformed into new rocks characteristic of the impact basins; special morphologies are created, such as impact cones with upwards

(continued)

Technical Box 5 (continued)
vertices and fragmented rocks known as impact breccias, which we can study on Earth (Fig. 3.7). The rock crushed by the impact is first thrown at great distances, creating a depression. Part of the evacuated material forms a rise at the edge of the nascent crater. Part of the material falls into the crater, thus forming an incoherent and crushed blanket. A fraction of the material can even leave the planetary surface if the attained speed is greater than the escape speed. This is how some very special meteorites, those belonging to the so-called SNC group, abandoned the surface of Mars, to strike the Earth some million years later.

Fig. 3.5 Orcus Patera is a huge crater (140 × 400 km) of ellipsoidal shape. Image Mars Express 204403. Mars Express ESA/DLR/FU Berlin (G. Neukum)

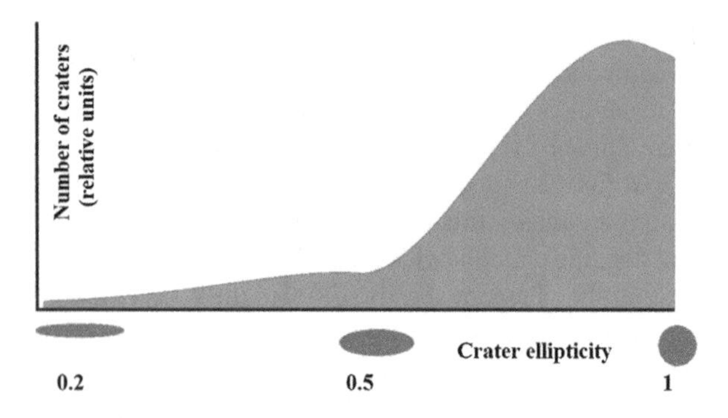

Fig. 3.6 Relative abundance of craters as a function of their ellipticity, i.e., the ratio between the minor to the major axis lengths. From Barlow (2015), modified and simplified. FVDB

Fig. 3.7 Breccia created by a meteoritic impact that occurred 650 million years ago. Gardnos (Norway). FVDB

Meteorite Craters on Earth

In 1980, an American group led by the physicist Walter Alvarez and his son Louis made a surprising discovery in the Italian town of Gubbio. A clay layer between the Cretaceous and Paleocene period dating back to 65 million years ago showed strong percentages of iridium, a very rare element on Earth, but not in metallic meteorites; its sudden and widespread abundance on Earth could be explained by the fall of a large meteorite onto our planet. The interesting thing is that the Cretaceous–Paleocene transition marks one of the greatest biological crises in geological history, which led to the simultaneous extinction of many plants and animals, including dinosaurs on the mainland and ammonites in the marine environment. The scenario was that an asteroid with a diameter of about 10 km plunged onto the Earth on the last day of the Mesozoic era. Releasing a huge amount of energy, the alien body caused immediate damage (explosion, heat wave, tsunamis in the case of impact on the ocean) and also long-term damage (shielding of solar radiation, fires), leading to the extinction of 60% of the species then living. However, the idea suffers from a lack of final proof: the meteorite-killer impact crater. There was no crater of the expected

Fig. 3.8 Left: The Barringer crater, one of the best-known examples of an impact crater. Right: The Ries Nordlingen in Germany, as seen by the radiometer ASTER on board the satellite Terra. Courtesy of the Jet Propulsion Laboratory. Left: image 126937593 FOTOLIA/nicolasdumeige, right: NASA image J. Allen, NASA/GSFC/METI/ERSDAC/JAROS, and U.S./Japan

dimensions (a few 100 km in diameter) and of the right age. Perhaps the meteorite had fallen into the ocean? A few years earlier, a meteorite crater 200 km in diameter had been discovered in the Yucatan Peninsula in Mexico by Alan Hildebrand. The geological observations about this crater, called Chicxulub, showed that it was probably the crater at the origin of the mass extinction that closed the Mesozoic era.

The big impact craters on Earth are much less numerous than those on Mars. Only geologically recent craters can still be seen in detail. Ten thousand years ago, a metallic meteorite 15–30 m in diameter stuck the North American surface at a speed of 15 km per second, creating the most well-known example of a meteorite crater on Earth: the Meteor Crater (or Barringer crater) 'Arizona' (Fig. 3.8). A crater that, though much smaller than Chixculub, is 1200 m wide and 200 m deep. There is an amusing story behind this crater. It was bought by an engineer named Barringer in the hope of finding a precious source of iron. Barringer was ridiculed, because, according to the ideas of the time, the crater was volcanic, and not the result of hypothetical bodies from space. He was correct about the extraterrestrial origin of his crater. But ironically, the research could not have succeeded, as we know today that the meteorite's body was completely disintegrated by the impact. The reason why the meteor crater is so well visible is simply because it is so very recent. Older astroblemes, though larger in size, are deleted in a short time. The Ries Nordlingen crater in Germany, for example, is not so obvious, despite the fact that it is only 15 million years old. Figure 3.8 highlights its presence only with the help of advanced techniques of remote sensing.

Gigantic Impact Basins on Mars

If most of Mars' craters are less than 100 km in diameter, there are also huge impact basins more than a 1000 km wide. The collision with Mars of three asteroids about 200 km in diameter about four billion years ago produced many huge recognizable basins: two, Hellas Planitia and Argyre Planitia, in the southern part of the planet and one, Utopia Planitia, in the northern (Fig. 2.1). Other smaller, but still enormous, craters are present both in the south and in the northern lowlands (the "QCD," see Chap. 2). While Argyre (−49°50′; 316° 41′ E) is a multi-ring basin 1500 km in diameter and "only" 5000 m deep (Fig. 3.3), Hellas (−42°42′; 70°00′E) is really amazing. This elliptical basin has an E-W-oriented axis diameter of 2300 km and 1900 km along the N–S direction. The whole of Western Europe would fit into Hellas. At the bottom of Hellas, the lowest point of Mars is reached at an altitude of about −8500 m. The walls are only a few degrees steep. Descending from the edges, we would reach the bottom of Hellas after a few 100 km, without realizing that we were going down. The sediments and lava flows covering the bottom of the impact basin are thicker in the center of the basin than on the edges, so that the shape of Hellas resembles the bottom of a bottle. The reason for this strange distribution of the material is not known. One of the most interesting things about these impact basins is the natural geometric tendency to accommodate anything that tends to go down. The bottom is thus filled with various sediments (watery and not), dust, and regolith. Impact basins are closed in, without emissaries, and thus if water enters, it could remain there for a while. In fact, there seems to be evidence that Hellas once hosted a lake. Also, some of the morphology inside Hellas appears to be clearly glacial (Chap. 4).

Regolith: The Secondary Product of Impacts

A crater 15 km in diameter typically has a depth of 1 km or a little more. In practice, it is a hole in the ground that has excavated nearly 200 km^3 of rock. If redistributed along the entire surface of the planet, the ejected material of this crater would drape it with a 2 mm thick layer. It seems a small amount, but there are thousands of craters of this size (in the northern hemisphere, the craters with a diameter between 14.1 and 20 km number about 1600[1]). Taking into account all of the impacts, the thickness of the fragmented rock ejected from the craters is on the order of many kilometers. All of this material, called regolith, is partially consolidated to form real sedimentary rocks and partly retained by the ice, while a smaller fraction remains free to wander as a fine powder, carried by the Martian winds.

3.2 Fractures

Fracture Patterns on Mars

As a consequence of the continuous horizontal movement, the tectonic plates of the Earth move at a speed of a few centimeters per year. If two continental plates collide,

[1]Barlow, N. G. (2015). Characteristics of impact craters in the northern hemisphere of Mars. *Geological Society of America Special Papers, 518*, SPE518-03.

initially, the rocks can withstand the tectonic stress, deforming. But at some critical stress, fracturing occurs. Fractures typically become arranged in planes lined up for long distances. In the case of compression of the lithosphere, inverse faulting is more common. Faults are called inverse when, due to compression stress of the crust, the two portions of the ground tend to "climb" over one another. There are also direct faults that form when the lithosphere is torn apart. Strike-slip faults do not involve shortening or opening, but rather a relative movement of the crust. Grabens are systems of direct faults parallel to each other, such as Tanzania's Rift Valley, a large, expanding area. A new ocean is emerging in the middle of the Rift Valley, which will, in ten million years or so, separate East Africa from the rest of the continent. Not all faults are a response to global structural phenomena. Many terrestrial faults are small and are due to localized phenomena, such as isostatic rebound of buoyant plutons (i.e., magmas lighter than the surrounding rocks) or local deformations.

If there has been some global tectonics on Mars, it was only of short duration compared to the planet's age, and therefore it is not possible to expect a structural geology similar to that of Earth. However, Martian fracture systems exist, although they seem to be more related to vertical rather than horizontal motion. As a result of the principle that, on Mars, geological activity and erosion is less efficient in erasing surface morphologies than on Earth, many different types of fracture have remained almost unchanged for billions of years, and so can be appreciated today.

Figure 3.9 shows the central part of the Alba Mons volcano, the most extensive volcano on Mars. Swarms of fractures directed south-north and continuing south (Ceraunius Fossae) and northeast (Tantalus Fossae) for 1000 km are interrupted by the volcanic building, which intercepted and modified the prevailing direction of the fractures from a linear to a circular pattern. This is because the volcanic building uphill pulls the crust towards the center of the volcano itself, which reacts by fracturing with this geometry. Something similar happens when a central floor expansion is underneath an asphalt pavement (Fig. 3.9, on the right).

From a global point of view, the western hemisphere of Mars is richer in fracture lines than the eastern hemisphere. The fractures especially affect the areas of the Tharsis bulge, Tempe Terra, Solis Planum, and Lunae Planum. An interesting observation is the following: fracture lines on Mars indicate an expansive, non-compressive style (more "direct" than "inverse" faults, according to the previous terminology). Swarms of related fractures often cross the planet along the west-east direction. Figure 3.10 shows a "patchwork" fracture that crosses a crater. It is part of a very extensive system of rifts in the Athabasca Valles area. All of this indicates that Martian tectonics is related to vertical or horizontal tension, and not to compressive movements of the crust.

Valles Marineris

Not all of Schiaparelli and Lowell's channels were optical illusions. Where the two scholars had noticed a large equatorial channel (called Coprates by Schiaparelli), in 1972 an east-west oriented canyon was recognized, now Valles Marineris.

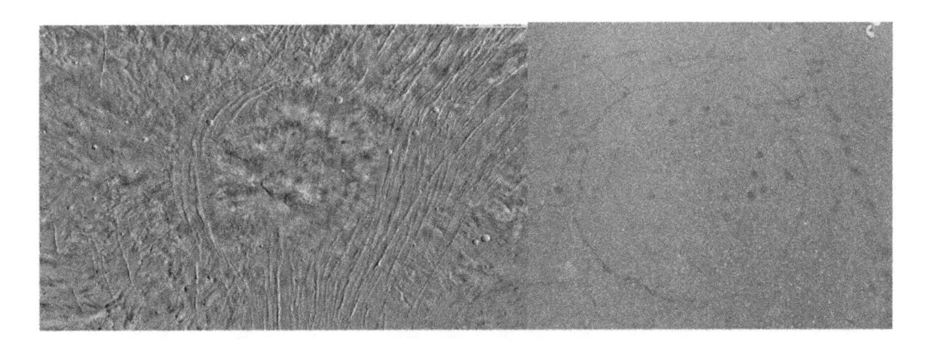

Fig. 3.9 Left: the central part of Alba Mons shows a system of concentric fractures (scene about 600 km across). Right: fracture system developed on concrete under which there was a central expanse of the soil (portrait size of the photo about 2 m). Left: THEMIS (Mars Odyssey), right: FVDB

Fig. 3.10 This crater in Athabasca has developed a central fracture, which is part of a system of 1000-km long fractures and a series of radial fractures perpendicular to the primary fracture. Flow-form morphologies are lava flows. Image THEMIS, V13300013, latitude 7.3388, longitude 161.372. The image width is 25 km. Courtesy of NASA/JPL-Caltech/Arizona State University. THEMIS (Mars Odyssey) NASA/JPL-Caltech/Arizona State University

Nowadays, only the mid-eastern part of Valles Marineris, the deepest portion of the valleys, is referred to as Coprates. Valles Marineris took its name from the Mariner 9 spacecraft that observed it for the first time. Over 3000 km long, it is a "canyon" wall system subdivided into several Chasmata, each up to 200 km wide and 5–8 km deep (Fig. 3.11).

Let us start with a short voyage through Valles Marineris, starting from the west: here, the valleys disappear into Noctis Labyrinthus, the night labyrinth, an intersecting semi-circular fracture pattern (Fig. 3.12). To the north, the fractures blend into smaller units, strongly oriented north-south. The labyrinthine structure is lost going east, where fractures acquire a prevailing east-west direction, the dominant

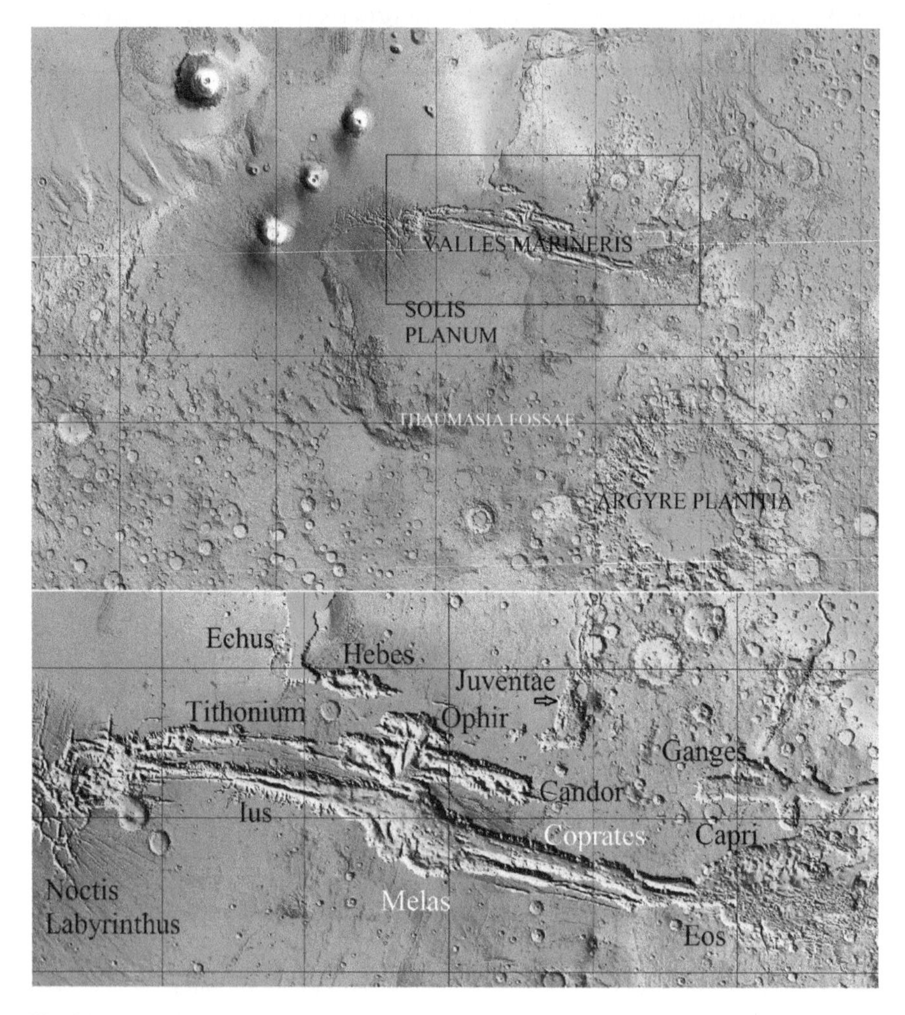

Fig. 3.11 Top: MOLA colorized map of the vast area surrounding Valles Marineris. Scene about 9000 km wide (at the latitude of Valles Marineris). Below, the enlarged area of Valles Marineris and its Chasmata. The image is about 3200 km wide. MOLA (MGS, NASA) mdf

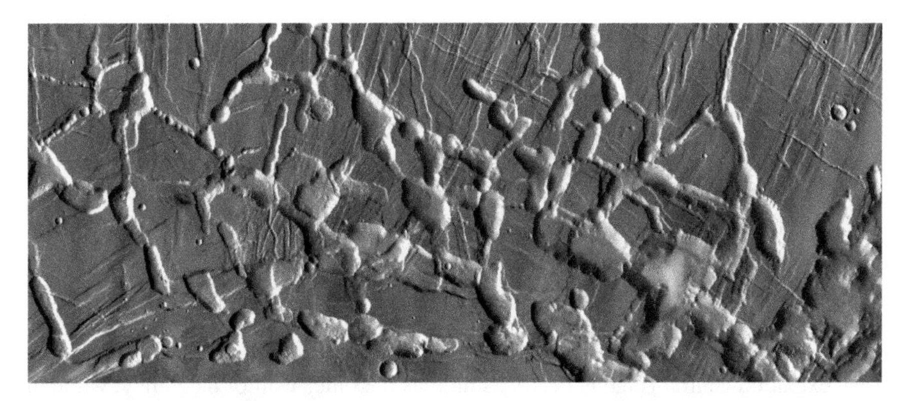

Fig. 3.12 A part of Noctis Labyrinthus. Notice the prevalent direction of fractures: towards the N-NE at the top. THEMIS daytime image. THEMIS (Mars Odyssey), NASA

Fig. 3.13 Candor Chasma. THEMIS, Courtesy NASA/JPL/Arizona State University, R. Luk

theme of Valles Marineris. We encounter an initial crossroads: in the northernmost portion, Noctis Labyrinthus blurs in Tithonium Chasma, a chasm without an eastern end. Further south, Ius Chasma is much longer and deeper, heading east to a wide area, Melas Chasma. The central part of Valles Marineris looks a bit like a Christmas Tree. From south to north, the names given were Candor Chasma (subdivided into east Candor and west Candor) and Ophir Chasma. Separated from the main valleys, but still considered part of Valles Marineris, there are also Echus, Hebes, and Juventae Chasmata. In the east, the Chasmata of Eos and Capri Chasma merge with the so-called chaos terrains and, through the channels heading north, Valles Marineris fades into the northern plains, crossing Chrise Planitia.

A view of Valles Marineris would look like Fig. 3.13. Desolate plains, yet rich in morphologies alternating with ridges. The canyon walls are at least 15°–35° steep.

We could climb to the top of the valley along a comfortable path. However, in many parts, especially the niches of ancient landslides, slopes sometimes exceed 45°. In fact, the walls have collapsed in many places in huge landslides: they can reach 1000 km^3, 50 times larger than the largest terrestrial landslides! Despite the fact that the Chasmata are filled with wall-leaked material, the bottom of Valles Marineris is fairly flat. Looking at pictures like those in Figs. 3.11 and 3.13, at least two details can be noticed immediately. Firstly, the bottom is filled with landslide deposits that have collapsed off of the walls. We will deal with them extensively. Secondly, there are ridges and central spines. The ridges can be examined well with topographic sections (Fig. 3.14).

In Hebes Chasma (Fig. 3.14, and the figure at the beginning of the chapter), one very extended central ridge is nearly as high as the outer wall of the valley, while in Melas Chasma and in most other locations, the ridge is lower. Huge swarms of fracture lines follow the profile of Valles Marineris, following them in parallel (Fig. 3.15, left). These are fractures resulting from tension. The presence of the valley depth of Melas Chasma creates a tendency of the rock wall to collapse towards the valley itself. Also, on Earth, fractures of this type are common, often announcing a new landslide (Fig. 3.15, right).

Valles Marineris has been compared to Arizona's Grand Canyon (Fig. 3.16), a comparison, as we will see, that is incorrect. However, it does have at least one thing in common with the American canyon. Valles Marineris is a natural excavation that allows us to examine its walls, and thus deduce a part of the geological history of this area of Mars. The rocks of the Grand Canyon show areas of rock outcrops (well visible in Fig. 3.16), alternated with porous portions where a talus has accumulated, i.e., broken rocks, sand and dust, recognizable by a gentler slope. This sort of rock in place-talus alternation is also well visible in the walls of Valles Marineris. In Fig. 3.17, a rocky layer in place forms a red strip (indicated by arrows) with high thermal inertia (indicating rock outcrops) alternated with an area of low thermal

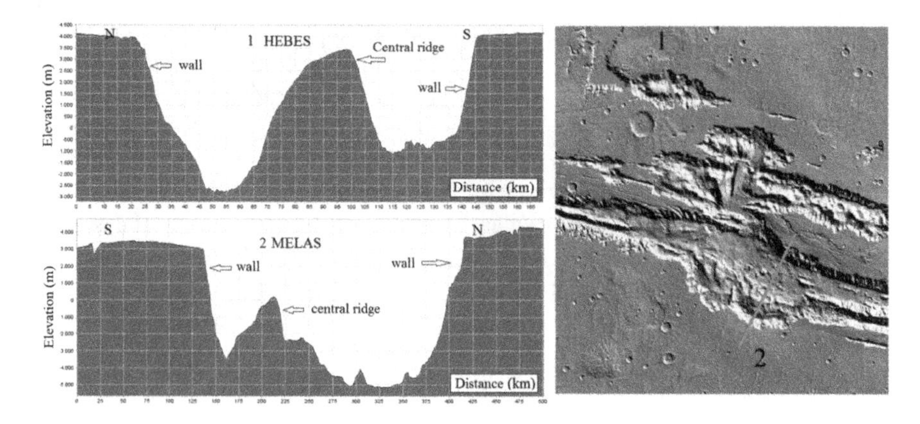

Fig. 3.14 Left: MOLA altitude sections through Hebes and Melas Chasmata. Right: the sections identified on a MOLA altimeter card. Vertical exaggeration makes the walls look steeper than they really are. MOLA (MSG), NASA, on JMARS platform, mdf

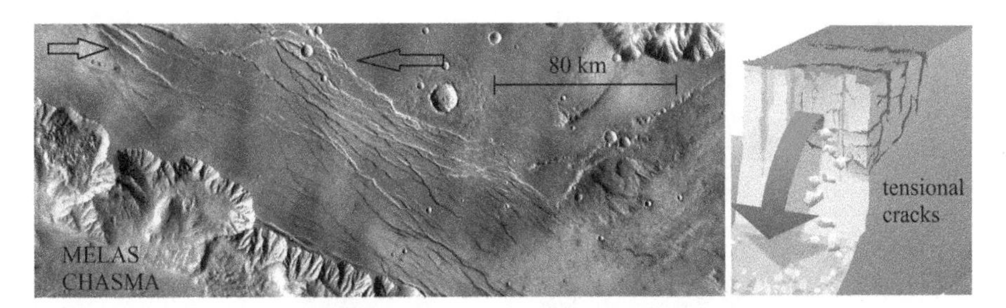

Fig. 3.15 Left: Long swarms of parallel fractures follow the outline of Valles Marineris, like here, north of Melas Chasma. The fractures form as a consequence of the huge tension that has accumulated due to the high relief of Valles Marineris walls. They are thus akin to the tensional cracks that are often observed at the top of unstable slopes as a sign of gravitative instability, sometimes culminating in a landslide. Left: THEMIS, right: FOTOLIA image 24419947, mdf

Fig. 3.16 Valles Marineris is sometimes compared to the Grand Canyon of Arizona, which is actually much smaller. In addition, the Grand Canyon was created by rapid river erosion (which is why its walls are very steep), while Valles Marineris has a different, still debated origin. Note the alternation of rocks in place and talus deposits, where the rock is more erodible. Image 70876790 FOTOLIA/Victoria Lipov

inertia (green), representing a talus (Technical Box 8). Downslope of the red layer, the thermal inertia is lower, indicating the presence of small blocks, and then it increases again. This geometry corresponds to a typical talus deposit, formed by the accumulation of fine material at the foot of the mountains, and also has correspondence in optical photos.

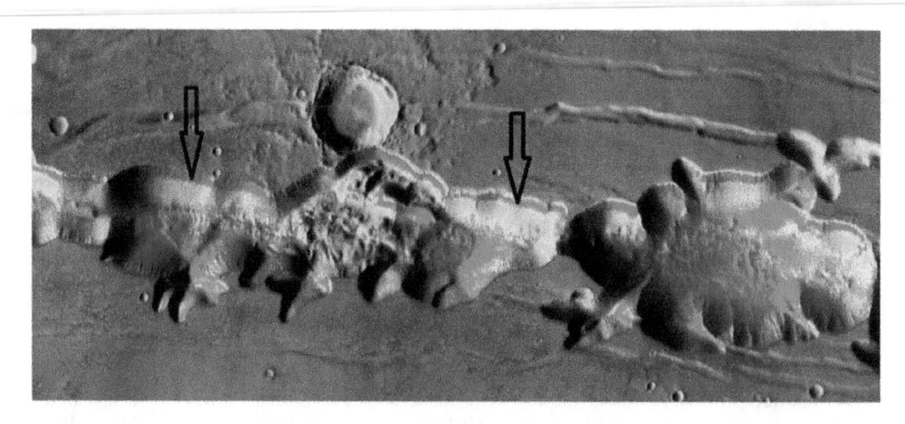

Fig. 3.17 High (red) and low thermal inertia (greenish) zones in Titonium Chasma. The red corresponds to values on the order of 600 J m^{-2} K^{-1} $s^{-1/2}$; the green, to about 60 J m^{-2} K^{-1} $s^{-1/2}$. The arrow shows a layer of rock. Since the walls' thermal inertia does not exceed 300 J m^{-2} K^{-1} $s^{-1/2}$, a layer of fine material probably covers the canyon walls. However, some stratifications show much higher thermal inertia (up to 2000 J m^{-2} K^{-1} $s^{-1/2}$). THEMIS (Mars Odyssey), NASA, mdf

FOURTH MYSTERY: How Did Valles Marineris Form?

It has already been mentioned how the analogy between Valles Marineris and the Grand Canyon of Arizona is fascinating but superficial. No river has eroded these amazing valleys as the Colorado did for the terrestrial canyon, though there may have been a lake in Valles Marineris. Another amazing difference is in the age: the opening of Valles Marineris has been dated to 3.5 billion years; the Grand Canyon dates to a mere 70 million years ago (this is, however, a record for this type of river feature, which is usually younger). A second terrestrial analogy is possible with the great African rift in Tanzania, a fracture thousands of kilometers long stretching from the Gulf of Aden to Mozambique, which has been active for just over 20 million years. The rift, which divides East Africa from the rest of the continent, provides a model for the fragmentation of Pangea, the supercontinent that existed at the beginning of the Mesozoic 200 million years ago and whose fragments correspond to today's continents. The separation of continents on Earth continued at a rate of several centimeters per year until, after hundreds of millions of years, the continents were separated apart, leaving submarine volcanoes that still exist in the middle of the Atlantic.

As we have seen earlier, the engine that moves the continents on Earth is internal heat. This engine decreases over time, partly because heavy elements no longer move to the center of the planet driven by gravity and partly because the amount of radioactive materials is running out over time. This is reflected in geological activity, which, on Earth, has decreased during the geological eras. As previously seen (Sect. 2.2), perhaps at the beginning of the geological history of Mars, the endogenous geological activity was sufficient to create a state of horizontal tension, which subsequently ceased with the decrease in the heat flow. According to this hypothesis, the horizontal currents opened the ancient Martian crust at Valles Marineris. The numerous landslides of Valles Marineris are by no means cut or distorted, though some are billions of years old. If Valles Marineris has opened slowly over time, should not we see traces of deformation of the oldest structures? Perhaps Valles Marineris formed very quickly? If Valles Marineris are the result of an aborted plate tectonics on Mars, the movement must have been fast at the beginning, with the opening rate then dropping about 3.5 billion years ago, the apparent age of Valles Marineris. The rift hypothesis, despite these problems, is not absurd and is preferred by many researchers.[2]

According to other experts, perhaps there was no opening, but rather a vertical sink at the bottom valley.[3] Also, on Earth, the vertical movements of continental masses had been hypothesized before the plate tectonics theory to explain the origin of mountain ranges and deep trenches. In the case of Valles Marineris, the collapse would take place for the withdrawal of a plutonic mass, i.e., a mass of deep magma. There is, however, an interesting detail: Valles Marineris occupies the center of the enormous Tharsis bulge 6000 km in length (see the altitude MOLA map at the beginning of the book). Is it possible that a swelling, and not a sinking mass, led to the formation of Valles Marineris[4]? An analogy with road pavement that is inflated by the roots of underlying plants is certainly very simplified, but also illuminating (Fig. 3.18). Another explanation for the origin of this quaint area is that the huge block of Thaumasia (Fig. 3.11) slid southward, ripping apart the lithosphere in correspondence with Valles Marineris.[5] If Valles Marineris is the result of fractures that widened with time (i.e., it is a tectonic valley in geological terminology), what was the relative movement of the blocks? Fig. 3.19 shows three possibilities. The drawing in (a) shows the initial opening. The topographic feature on the right precedes the opening, and may provide information on the geometry. (b) shows a way in which a tectonic valley might open up like a rift, where normal faults prevail. (c) and (d) show strike-slip faults, i.e., parallel to the axis of the initial opening. If there has been a movement to the right (c) or left (d), it can be deduced from the final position of the geological structure.

It has been suggested that Valles Marineris is a system of tectonic valleys with left strike-slip fault displacement of about 160 km.[6] However, this is difficult to assess

---

[2]Schultz, R. A. (1991). Structural development of Coprates Chasma and western Ophir Planum, Valles Marineris Rift, Mars. *Journal of Geophysical Research: Planets*, *96*(E5), 22777–22792.

[3]McCauley, J. F., Carr, M. H., Cutts, J. A., Hartmann, W. K., Masursky, H., Milton, D. J., & Wilhelms, D. E. (1972). Preliminary Mariner 9 report on the geology of Mars. *Icarus*, *17*(2), 289–327; Tanaka, K. L., & Golombek, M. P. (1989). Martian tension fractures and the formation of grabens and collapse features at Valles Marineris. In *Lunar and Planetary Science Conference Proceedings* (Vol. 19, pp. 383–396).

[4]Lucchitta, B. K., McEwen, A. S., Clow, G. D., Geissler, P. E., Singer, R. B., Schultz, R. A., & Squyres, S. W. (1992). The canyon system on Mars. *Mars*. The University of Arizona Press, Tucson (USA). 453–492.

[5]Anguita, F., Farelo, A. F., López, V., Mas, C., Muñoz-Espadas, M. J., Márquez, Á., & Ruiz, J. (2001). Tharsis dome, Mars: New evidence for Noachian-Hesperian thick-skin and Amazonian thin-skin tectonics. *Journal of Geophysical Research: Planets*, *106*(E4), 7577–7589.

[6]Yin, A. (2012). Structural analysis of the Valles Marineris fault zone: Possible evidence for large-scale strike-slip faulting on Mars. *Lithosphere*, *4*(4), 286–330.

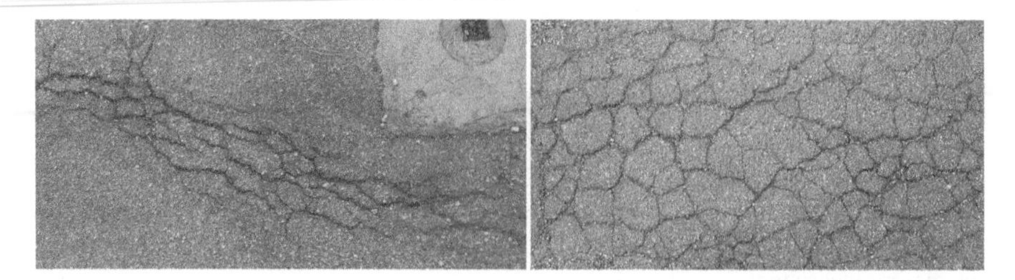

Fig. 3.18 The fragmentation of road cement caused by the expansion of underlying tree roots provides a simple analogy with the fracture system of the Noctis Labyrinthus-Valles Marineris area. Left: The patterns of linear fractures due to roots that extend from left to top right down resemble the setting of the axis of Valles Marineris. To the right: A more uniform expansion caused by a jumbled system of many roots underneath produces a fracture system that calls to mind Noctis Labyrinthus. FVDB

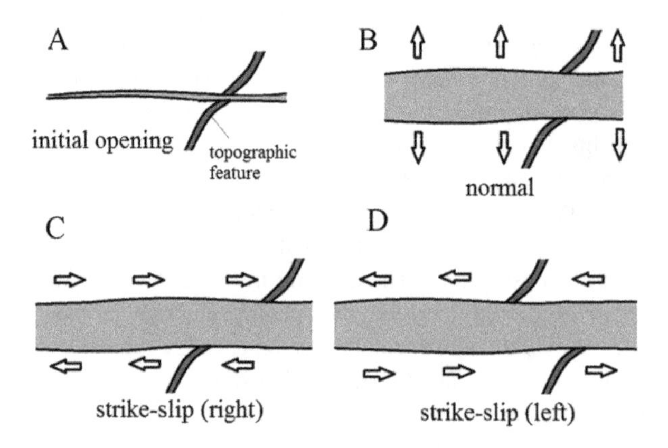

Fig. 3.19 Possible fault systems in Valles Marineris. FVDB

with precision, because the displacement of the geological structures predating the opening are not so apparent. It is also difficult to understand how the Noctis Labyrinthus area may have remained blocked with regard to this hypothetical movement. As demonstrated by the presence of the ridges, Valles Marineris appears mostly to be controlled by a system of normal faults (b in Fig. 3.19), thus reflecting an expansion of the area. It has been suggested[7] that the impact that gave rise to the large basin of Argyre Planitia southeast of Valles Marineris had a dramatic consequence at a distance of at least 10,000 km (Fig. 3.11, top). According to this hypothesis, the impact set the lithosphere in tension, which began a horizontal

[7]Yin, A. (2012). An episodic slab-rollback model for the origin of the Tharsis rise on Mars: Implications for initiation of local plate subduction and final unification of a kinematically linked global plate-tectonic network on Earth. *Lithosphere*, *4*(6), 553–593.

movement: a sort of Martian plate tectonics, triggered by the asteroid impact. The Thaumasia block (Solis Planum) began to move south, and part of the lithosphere was subducted in correspondence with the Thaumasia Fossae. To the north of Solis Planum, the Martian crust was torn apart, forming Valles Marineris. However, this hypothesis suffers from a series of problems. Firstly, no type of global tectonics is known to result from an impact. In addition, Hellas Planitia, an even bigger catchment area than Argyre, has not generated any kind of valley system.

According to another view,[8] the story of Valles Marineris began north, and not south, of the valleys. The transition from the southern mountainous regions to the northern lowlands occurs, as we have seen, along a line. Since north of this Martian dichotomy line, elevations are lower than in the south, there is a mechanical traction that would tend to make the crust of the southern mountains slip to the north. Without further forces, however, there is no movement, because the traction force is too low and the rock resistance stabilizes the whole area. However, when Tharsis's swelling was formed, the pulling force resulting from that swelling added to the one already present due to the presence of the global dichotomy. Thus, throughout that area, the tectonic stress grew sufficiently high to make the crust slide to the north.

If the initial formation of Valles Marineris is the subject of controversy, there is little doubt that the widening of landslides has been very important in its later evolution, as we shall see in the next section.

3.3 Catastrophic Landslides

Landslides are common on Earth. Often caused by earthquakes, reduced support to the base or the presence of percolating water, they can bring nasty consequences. Rapid landslides are roughly classified into three families: debris flows, composed of wet clay, sand and large blocks. Falls, generally small in size and taking place over a short distance. And finally, rock avalanches, derived from the failure of a rock mass larger than ten million cubic meters. Rock avalanches are mobile mass flows, and their capability to travel horizontally increases with their volume. When exceeding 10 km^3 of material (only a few examples of such large landslides are known on Earth), rock avalanches can travel at very high speeds and reach distances of over 10 km.

Valles Marineris's largest landslides often reach volumes of several 100 km^3. On Earth, one of the largest landslides in the world, the Saidmarreh rock avalanche in Iran, is "only" about 10–20 km^3 large. Thus, the large Martian landslides may be fifty times larger than the Saidmarreh landslides. And these are minuscule in

[8]Andrews-Hanna, J. C. (2012). The formation of Valles Marineris: 3. Trough formation through super-isostasy, stress, sedimentation, and subsidence. *Journal of Geophysical Research: Planets, 117*(E6). Andrews-Hanna, J. C., & Lewis, K. W. (2011). Early Mars hydrology: 2. Hydrological evolution in the Noachian and Hesperian epochs. *Journal of Geophysical Research: Planets, 116* (E2); Andrews-Hanna, J. C. (2012). The formation of Valles Marineris: 1. Tectonic architecture and the relative roles of extension and subsidence. *Journal of Geophysical Research: Planets, 117*(E3).

comparison to the landslides from Olympus Mons, the largest of which have totaled one million cubic kilometers (Sixth Mystery). One of the reasons for the huge volume shown by Martian landslides has to do with the reduced gravity. The smaller gravity field on Mars compared to that of our own planet allows larger masses to accumulate before the threshold for slope instability is reached. Some of the landslides have gone a long way, reaching distances of up to nearly a 100 km. Terrestrial records once again appear modest compared to the Martian ones: Saidmarreh has reached a distance of only 20 km. Many large landslides in Valles Marineris have reached distances five times as great!

Figure 3.20 shows a massive landslide in Melas Chasma. The morphology of the landslide closely resembles terrestrial landslides: the semicircular niche, partially covered by fragmented landslide material that has failed to get away. Below, a landslide slope plain. After reaching the almost flat base of Valles Marineris, the material did not stop at all. On the contrary, it is in this flat area that the landslide has shown the largest mobility, splashing away over tens of kilometers at high speed. We know that the speeds were high (in some cases, over 300 km per hour) because the landslide managed to pass over a number of mounds on the ground. The principle that permits velocity estimates is the same as that regarding a skier who, after reaching the top speed at the foot of a slope, wants to ascend to the top of a small mound. Evidently, he needs to have gained minimum speed to travel uphill for a while and reach the top. By measuring the height reached by the skier, one can calculate his speed before ascending up the mound. The mound indicated in Fig. 3.20 with the number "4," for example, is nearly 1000 m high. A simple kinematic calculation shows that the landslide must have had speeds on the order of 300 km per hour or more to be able to pass over it. The much higher hill marked "5" has not been surpassed. From the measured height, the velocity was on the order of 400 km per hour. Computer simulations also show the possibility that a very large landslide collapsing from a 9 km high slope with the properties of Valles Marineris may reach the astounding velocity (at peak) of 700 km per hour! Although this is the case for a very mobile landslide, undoubtedly, these rock avalanches were extremely fast.

What caused the landslides of Valles Marineris? Fig. 3.21 shows a landslide in Noctis Labirynthus. An impact crater appears just at the top of the landslide, suggesting that a meteorite fall may be the culprit. In fact, we have seen that the fall of a meteorite releases a huge amount of energy. A meteorite 1 km in diameter and with a density of 7000 Kg/m^3, falling on the surface of Mars, releases an energy of 10^{20} joule and digs a crater of about 20 km. The energy of a landslide like the one in the figure is one hundredth the energy released by the impact. So, there is plenty of energy available from the impact. However, many landslides in Valles Marineris do not appear to be related to any impact crater. Experience gained with terrestrial landslides shows that a deep and sharp canyon like Valles Marineris is prone to instability, irrespective of a trigger like meteorite impact. Computer simulations show that many locations within Valles Marineris may fail one day or another, especially if the rock is not intact, but fragmented. The linear fractures in Fig. 3.15 parallel to the slope outline precisely show an unstable situation, which, on our planet, would elicit much concern. We can be fairly certain that more landslides will collapse there again, even though it may take millions of years!

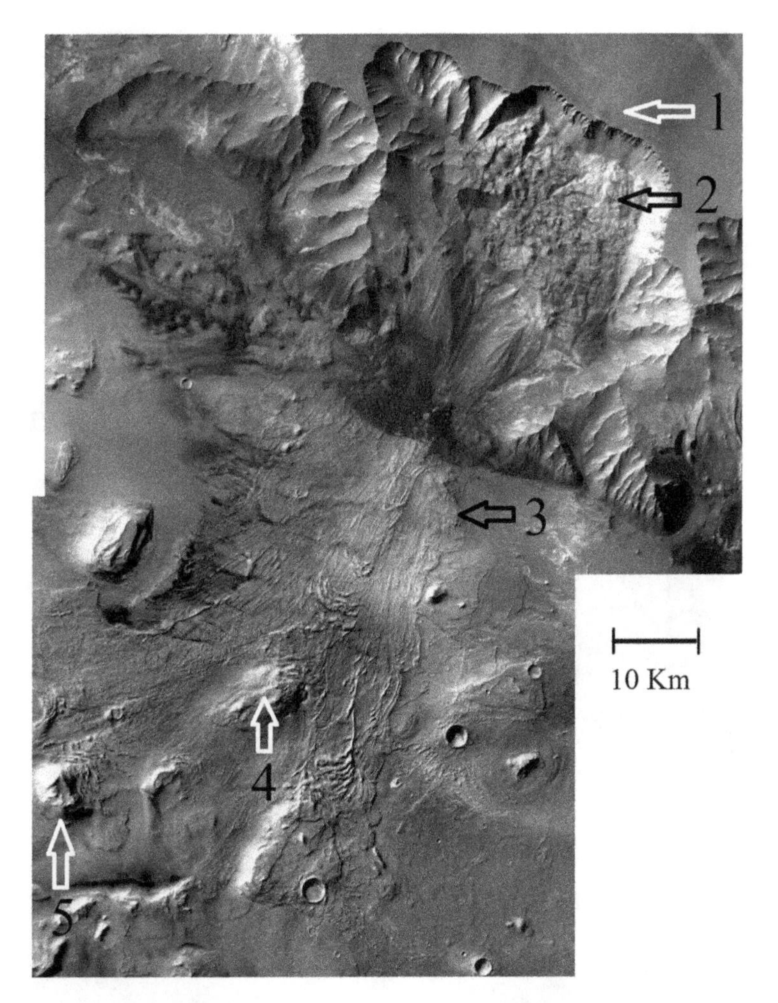

Fig. 3.20 The steep edges of the Canyon of Valles Marineris have collapsed in many places. The figure shows one landslide in Melas Chasma of Valles Marineris. Note the amphitheater-shaped detachment niche (1), the collapsed material in the niche (2), the horizontal spread in the bottom of Melas Chasma (3), and the impact with several mounds (4 and 5, although many others are evident). This particular form of collapse indicates homogeneous soil and is common to many terrestrial landslides. Valles Marineris's landslides reach distances of many tens of kilometers, despite the fact that the base of the valley is almost horizontal. CTX (MRO), NASA, mdf

Another enigmatic feature is the reason why Mars' landslides travel so far. A clue as to landslide mobility is provided by comparing the horizontal spread of the landslide (called the "runout") with the vertical fall. On Earth, a very small landslide of dry rock typically travels horizontally twice to three times its vertical fall. However, large terrestrial landslides (larger than 500 million cubic meters) may travel six or more times their vertical fall, and even larger ones travel ten times their vertical fall. This behavior is bizarre and poorly understood, as we know that dry

Fig. 3.21 Looking at Noctis Labyrynthus from east to west. A large 25-mile long landslide can be noted in the foreground. Could it have been triggered by the impact of the meteorite that left a crater right at the point of origin of the landslide? Most of the Valles Marineris landslides, however, do not appear to be related to impact craters. Courtesy of NASA/JPL/Arizona State University, Coordinates: −9.94, −95.44. THEMIS (Mars Odyssey), Courtesy NASA/JPL/Arizona State University

granular avalanches are scale-invariant. This means that, as both theory and experiments with granular materials show, the ratio between runout and fall height should always be the same, irrespective of the scale, which is in blatant contrast with observations. Similar to their terrestrial counterparts, Martian landslides are very long considering their fall height, and this extra mobility increases with the volume. They may even travel twenty times their fall height! We shall come back with a possible explanation in Chap. 4.

Landslides are not limited to Valles Marineris. Other smaller landslides are distributed across the Martian surface, and particularly at the rims of craters in the southern hemisphere. A special type of landslide is shown in Fig. 3.22. Picturesque strips of dark material a few kilometers long appear on the brighter background, stretching along the direction of the maximum slope. Often called recurrent slope lineae, these are very superficial movements that involve only the top layer of regolith. Moving downwards, the sandy avalanche has collected more loose material, and the whole collapse has resulted in these black lobes (Fig. 3.22). A debate exists as to the wet versus dry nature of these superficial collapses. Anything that may involve past water on Mars (and especially present, since these collapses are recent) is particularly important. We know from terrestrial studies that the more the material is cohesive (cohesion of loose sand implies water), the more the avalanche opens up, developing a triangular-shaped failure surface (Fig. 3.23). It therefore seems that the Martian figures in Fig. 3.22, more festoon-shaped than triangular, have been dry. This is confirmed by a recent research showing that, irrespective of their size, such superficial flows all collapse at the same angle. This is characteristic of dry flows, the angle of collapse being coincident with the angle of repose of the regolith.[9]

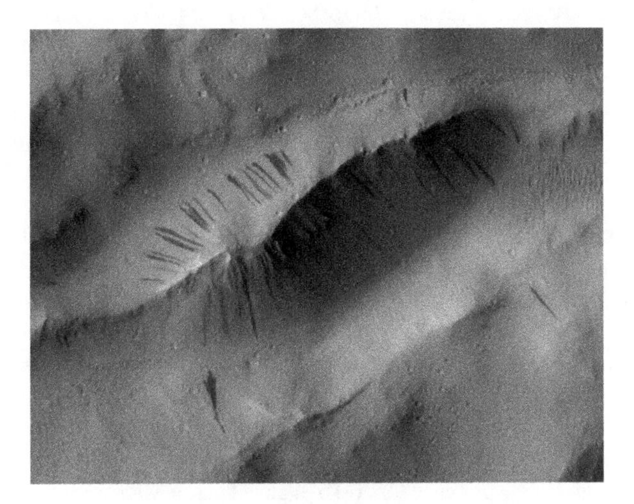

Fig. 3.22 Surface landslides appear as black stripes on the light regolith. They start from a point at the top of a hill, in this case, the ripples of the north aureole of Olympus Mons. Width of the image from left to right of about 6 km. CTX image G19_025602_2088_XN_28N135W. CTX (MRO), NASA

[9]Dundas, C. M., McEwen, A. S., Chojnacki, M., Milazzo, M. P., Byrne, S., McElwaine, J. N., & Urso, A. (2017). Granular flows at recurring slope lineae on Mars indicate a limited role for liquid water. *Nature Geoscience*, *10*(12), 903.

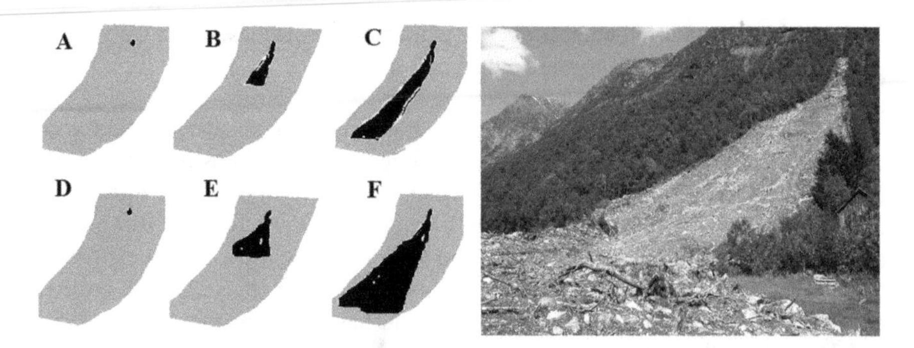

Fig. 3.23 Formation of Dark Stripes on Mars. (**a**): Instability of the regolith at a high point. (**b**): The white veneer avalanches, revealing the darker soil beneath. The movement makes the parts more unstable in a chain process. (**c**): Finally, the mass stops at low angles. (**d** and **e**) shows the same process when the material is cohesive. In this case, a triangular shape is developed and not a festoon-shaped landslide. Right: example of cohesive failure on colluvium (cohesive soil) in southern Norway. Note the significantly widening triangular shape. FVDB

Technical Box 6: Geological Survey on Mars

Formations are the basic lithostratigraphic units used in terrestrial geology. One formation includes an assembly of sedimentary, igneous, or metamorphic rocks (the same formation can embrace different rocks). Some formations have been deposited in lacustrine or marine environments before turning into rock, while others represent a portion of the earth's crust subjected to such intense pressure and temperature that new minerals and rocks have been created in the metamorphic process. From field studies, picturesque geological charts are constructed, in which a color represents a formation. The description of the first geological formations dates back to the times when the hammer was practically the only tool available to the geologist. In regard to Mars, we do not have the option of examining the rocks so closely. Martian geology is therefore studied with different techniques. Spacecraft provide wide-field or more detailed images at different light wavelengths, from which mineralogical composition can be inferred. Moreover, landers and rovers have also analyzed rocks from a limited number of locations. It has been found that the distribution of minerals and rocks is much more uniform on Mars than on Earth. The rocks are, in fact, all eruptive or sedimentary, but the latter also derive from the same eruptive rocks broken by various processes (primarily meteorite impacts), reworked, and then redeposited in the presence (or not) of water. Thus, a Martian geological map is less detailed than one of Earth, partly due to a lack of information at the outcrop level, but also due to the scarcity of mineralogical and petrographic variety.

A terrestrial geological map also holds information on the age of the formations, obtained either by absolute or relative methods. The latter include the study of fossil content in sedimentary rocks. Not so on Mars, for which dating is much more indirect (Technical Box 3). However, the knowledge gained through spacecraft and rovers has permitted creation of a division of Martian formations based on the nature of the rocks and their age, from which a geological map of Mars has been built. As an example, Fig. 3.24 shows a geological map of Mars elaborated by the USGS

(continued)

Technical Box 6 (continued)

(United States Geological Survey). This portion comprises part of the volcanic area of Tharsis and Valles Marineris in two interpretations: an initial image based on Viking data and a much more recent map. Note the emphasis on the level curves in the former and the more refined outlines in the latter. The interested individual should print the geological charts with a color plotter and consult the leaflet.[10]

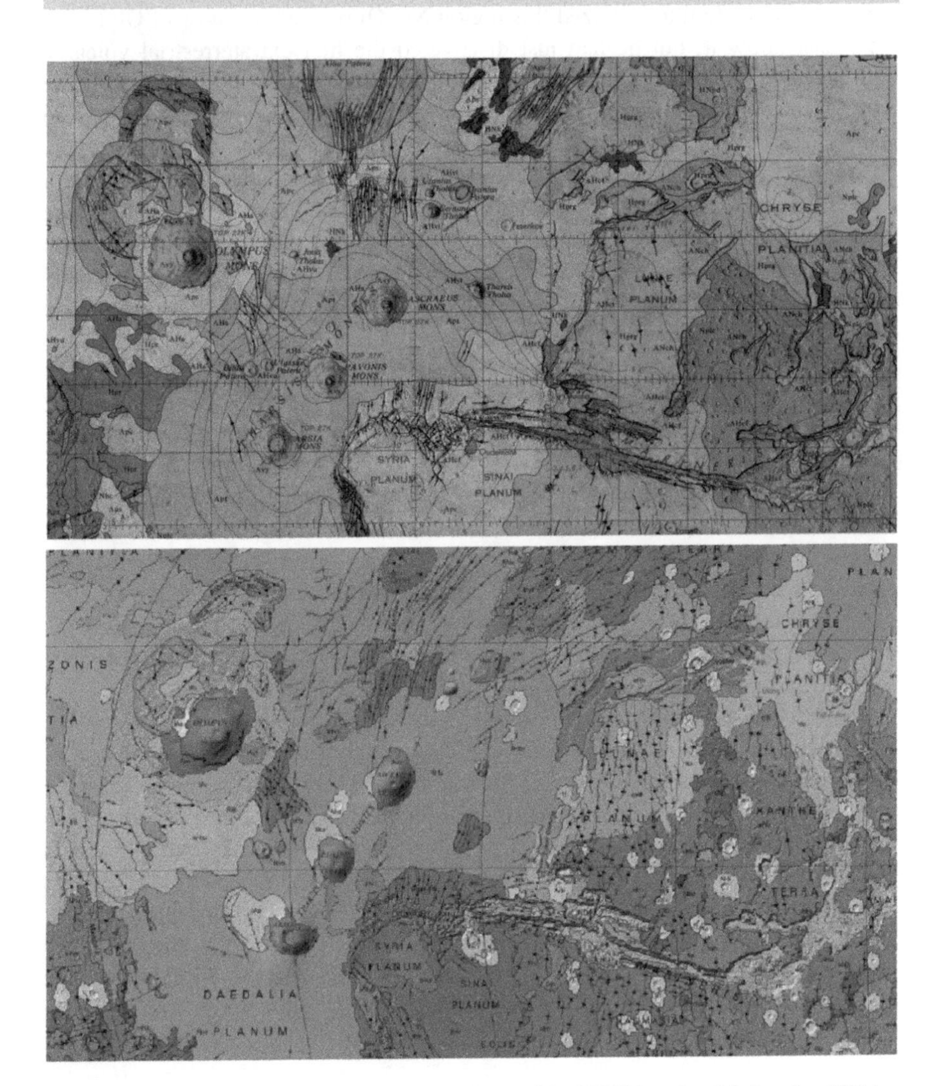

Fig. 3.24 Top: part of the first Martian geological map of the USGS based on Mariner 9 and Viking data. This portion is centered around the volcanoes of Tharsis and Valles Marineris. Bottom: A modern geological map of the same area. Courtesy of USGS

[10]Map and leaflet by USGS (Ken Tanaka and collaborators) can be retrieved at https://pubs.usgs.gov/sim/3292/

3.4 The Volcanoes of Mars

Olympus Mons and Other Huge Volcanoes

During the opposition of 1877, Schiaparelli discovered a white spot on the surface of Mars similar to the ice cap of the Martian poles. Being far from the poles, according to Schiaparelli, the spot could only have one explanation: snow on a high mountain. He thus baptized this region Nix Olimpica, the snow of Olympus. He didn't know it, but he had just discovered the first extraterrestrial volcanic building: a huge shield volcano, re-named Olympus Mons in the 1960s. It is the highest volcano in the solar system and has a number of features that are hard to believe.

Six hundred kilometers in diameter, over 22 km tall, a volume equal to 6 million cubic kilometers, and an extension equivalent to that of France (Fig. 3.25). If we ascended up the slopes of Olympus Mons from the base to the top starting from the flat Amazonis Planitia area 500 km west of the volcano, we would not see any mountain on the horizon. Only dust and stones on the ground would lead the eye to the distance, immersed in a panorama similar to that of Utopia Planitia, visited by the Viking 2 lander. We would then encounter a series of parallel corrugations up to hundreds of meters in length forming long-distance ropes. Walking along the monotonous landscape, after several days, Olympus Mons would still be invisible, far beyond the horizon.

Such corrugations are one of the most enigmatic features of Olympus Mons. Called aureoles, these are strange lobed structures that surround the volcano from all directions, but mostly from the north and the west (Fig. 3.25). To the northwest, they reach a distance of over 700 km from the edges of the volcanic building. It is obvious that such aureoles radiated from Olympus Mons. But what are they exactly? How did they propagate along a flat area crossing a distance equal to that between Paris and Berlin? (Sixth Mystery.)

At a distance of 400 km, Olympus Mons would be completely beyond the horizon. Coming closer along the aureoles, at a distance of 200 km, the summit of the volcano would appear first as a perturbation of the line on the horizon. Then, we would hike up the basal scarp, well visible in Fig. 3.25. This would really take us up the main volcanic edifice: a slant of about 30°—easy for a walker, but dotted with blocks of all sizes. In fact, a unique case for the volcanoes of Mars, Olympus Mons is separated from the base by this basal scarp, a steep wall covered in some places by recent lava flows. This steep slope is enigmatic, too (Fifth Mystery). Then, once we had reached the plateau of the shield volcano after a rise of about 8 km (an altitude difference equal to that between the sea level and Mount Everest), we would be able to perceive its majesty, without seeing it all in full. And finally, once we had reached

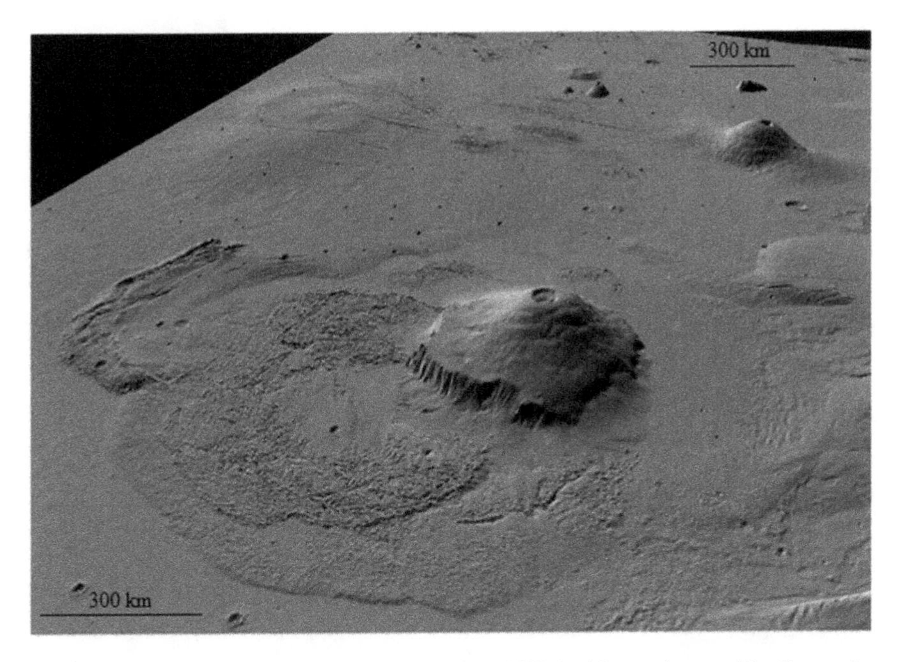

Fig. 3.25 The surroundings of Olympus Mons in a MOLA altimeter image. Alba Patera, the broadest volcano on Mars, is to the left in the background. The conical edifice on the right is Ascreous Mons. In the center, Olympus Mons, with its aureole (halo). The prospective lines lead in the East-Northeast direction. The half-moon feature on the left merging with the halo is Acheron Dorsum, with its trenches (Fossae). MOLA (MGS), NASA, on JMARS platform

the volcanic plain, the inclination would become smaller and the walking easier. We would walk on more recent lava flows that have built up the volcano, but 300 km would still separate us from the caldera.

The enormous height of the volcano also locally altered the atmospheric currents. Just as with water vapor over a terrestrial mountain chain, the adiabatic expansion of the air forced to move upwards condenses the carbon dioxide, creating thin clouds at the top of the volcano. The resulting haze, and not snow, forms the whitish layer sometimes visible at the top, and which revealed the presence of a very tall mountain through Schiaparelli's eyepiece. Perhaps the process went on for hundreds of millions of years, maybe billions. In fact, towards the western direction at the foot of Olympus Mons and other volcanoes in the Tharsis area, there appear strange lobate deposits. These are traces of ancient glaciers or high "rock glaciers" similar to terrestrial ones, indicating ice made from water caught in the atmosphere at the volcanic building (Chap. 4).

Distribution of Martian Volcanoes

Numerous other volcanoes bear witness to effusive activity on Mars, which has continued for billions of years. Compared to their terrestrial counterparts, the Martian volcanoes are often much larger and taller. The volcanoes are not isolated, but mostly present in four distinct areas: the Tharsis region, Elysium, the Syrtis Major, and around (and north-west of) Hellas Planitia (Fig. 3.26). There are also small volcanoes distributed here and there on the surface of the planet. Let us look at the main ones.

Tharsis is the region with the largest volcanoes (Fig. 3.27). The widest is Alba Mons (formerly Alba Patera), a huge plateau 1100 km in diameter and 6 km in height (Fig. 3.9). In the center lies a caldera some 10 km in diameter, surrounded by a plateau of about 300 km. At greater distances, the volcano becomes steeper (although it is less than half a degree steep everywhere). It erupted in very long lava flows: at distances of 1400 km from the crater, lava from Alba can still be found. The edges of such flows give the periphery of Alba Mons a mushroom-like appearance. The presence of a series of southern fractures oriented in the N-S direction, continuing NE with centrifugal orientation, and the circular fractures in the central volcano region were shown earlier (Fig. 3.9). Such strange concentric fractures recall

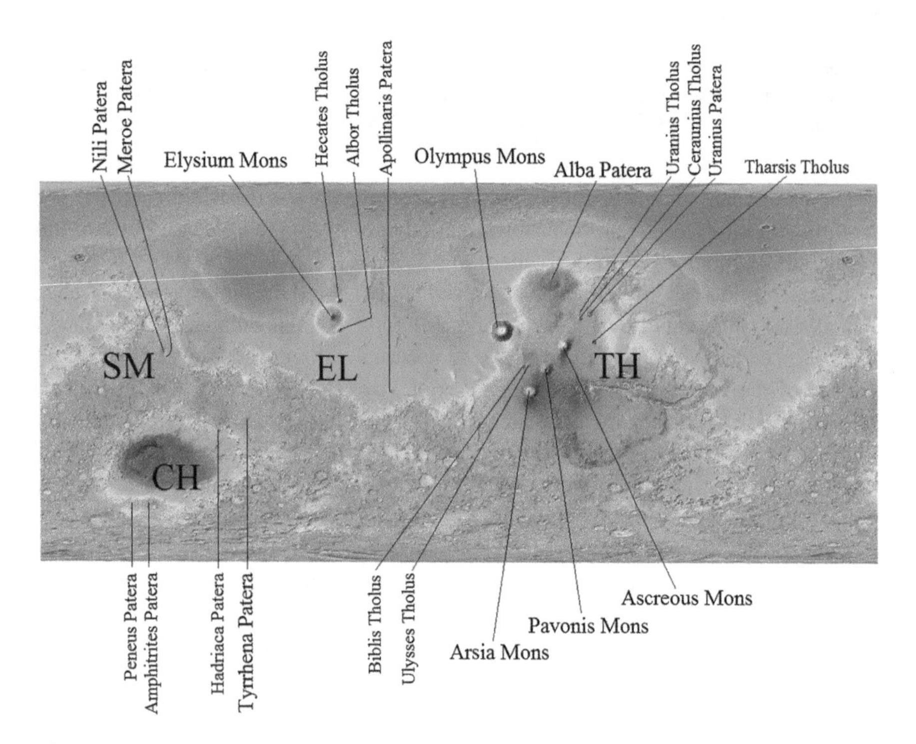

Fig. 3.26 The four volcanic provinces of Mars and their main volcanoes. SM: Syrtis Major; CH: Circum-Hellas; EL: Elysium; TH: Tharsis. MOLA (MGS), NASA map, mdf

Fig. 3.27 MOLA map of the principal volcanoes of the Tharsis region. MOLA (MGS), NASA map, mdf

certain Venusian volcanoes. Lava flows are of two types: channeled or in the form of a sheet flow. The distances of hundreds of kilometers reached on slopes below 1° show an exceptional fluidity of the lava, which, on Earth, is common only for basaltic lava. To reach these proportions, Alba probably began to erupt at least two or three billion years ago, though many of its flows appear much fresher.

To the southeast of Olympus Mons, there appear three large volcanoes arranged along a line. These are Arsia Mons, Pavonis Mons and Ascreous Mons, collectively called the Tharsis Mountains. Arsia Mons, the southernmost volcano in the province of Tharsis, is nearly 18 km high and has a diameter of 400 km. The overlapping of later, coarser collapsing deposits in the east and west is clearly visible. Pavonis Mons, a little smaller, also has well visible lava streams, but also concentric fractures (perhaps originally more extensive, but now covered by the lava of the volcano itself) and, as for Arsia Mons, it is also covered with fine material. Ascreous Mons, a bit bigger, is also endowed with a constellation of lava flows. Along the extension to the NE of the Tharsis Mountain line, there appear smaller volcanic buildings: Uranius Tholus, Ceraunius Tholus and Uranius Patera, and further south, a triangular volcano known as Tharsis Tolus. Ceraunius Tholus is almost completely unique, because of its many radial channels, probably dug by running water (Fig. 3.28). Why are the radial channels so neat on this volcano and not on others? Perhaps Ceraunius is a heap of tephra (pyroclastic material such as ash and lapilli), which represents much more incoherent and erodible materials than a lava flow.

The second volcanic province of Mars is known as Elysium (Fig. 3.26). It includes a number of volcanoes far inferior to Tharsis, but given its distance from Tharsis, it should be considered as a distinct province. Elysium Mons, a volcano of

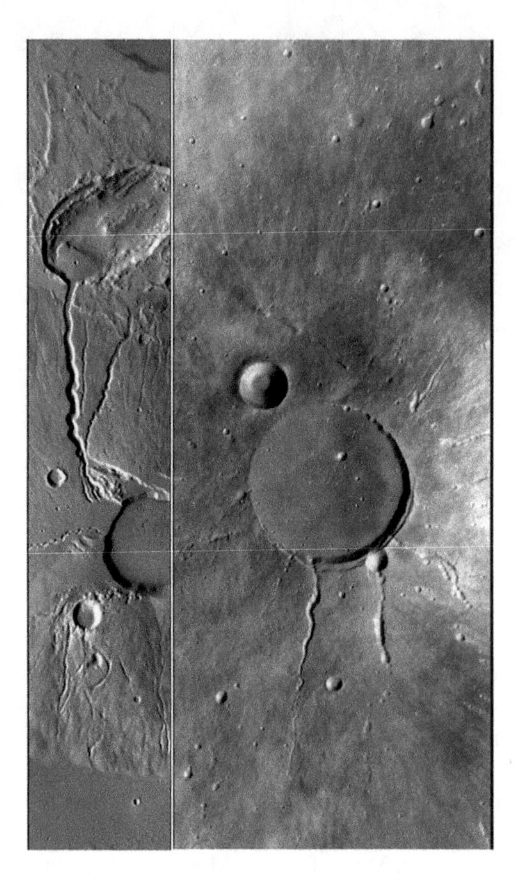

Fig. 3.28 Left: the caldera of Ceraunius Tholus and the material that forms the sides of the volcano were dug out of water. A series of torrents from the upper part of the volcano descended to form a lake in the ellipsoidal impact basin visible high up in the image. The erodible material was probably made up of poorly consolidated pyroclastic rocks. Image CTX B04_011399_2045_XN_24N097W. Image width from left to right about 30 km. Right: the summit of Elysium Mons. Note also the radial fracture system and the number of tiny craters, even inside the larger crater. Image CTX B22_018247_2048_XI_24N213W. CTX (MRO), NASA

huge extension (some of its lava flows reach 1700 km from the central eruptive vent!) is a very flat shield, "only" 5 km high and sloping less than 1° (Fig. 3.27). Hecates Tholus, much smaller, has neater borders and rather steep sides (30°); it seems to be partially embedded in the immense flows of nearby Elysium Mons. Many of the lava streams that characterize Hecates come from eruptive mouths now buried by subsequent magmas. Albor Tholus, located southeast of Elysium Mons, exhibits a series of concentric fractures at its southern edge.

A series of enigmatic, probably very ancient volcanic centers are concentrated around the Hellas Planitia area, the huge impact basin on the southern hemisphere of the planet. The most notable is the 400 km wide, elongated Hadriaca Patera. The interesting thing about this volcano is that it shows gullies incised along its flanks by water, indicating a consistency more similar to volcanic ash than hard lava flows. Hadriaca is relatively old, dating back 2–3 billion years ago. The fourth region is that of Sirtis Mayor, a dark area among the first to be recognized in the telescopes of the eighteenth century (Chap. 1). Despite the telescopic accomplishment, the two volcanoes of the Syrtis region, Mereo and Nili Paterae, were recognized as such only in the 1980s. We now know that the darkness is precisely due to the presence of the two volcanic vents, because these two volcanoes emit mafic lava particularly rich in dark minerals.

Volcanism on Earth and Mars

Volcanism is a common phenomenon on Earth, and many human communities have witnessed devastating eruptions. Sometimes, a volcanic eruption is mild and gives rise to lava flows that descend from the sides of the volcano, stopping a few kilometers from the eruptive mouth. Other times, the eruptions are explosive and cause real catastrophes, such as during the eruption of Krakatau (1883) or the more recent one of St. Helens (1980). What does the variety of volcanic phenomena depend on? And what are the differences with the volcanoes of Mars? Despite volcanic risk being seen as almost an anomaly, volcanism is, in fact, a normal geological phenomenon, not only on Earth but on many other planets and moons of the solar system.

Terrestrial volcanoes are not distributed randomly. Many volcanoes concentrate along the edges of the lithospheric plates, those areas where the crust and upper mantle of the Earth are very fractured. The areas between the plates are weak areas, and there, the magma from the depths finds an easy way along the crust. Also, the relative movement of the plates continually changes the pressure at some depth; often, melting of the upper mantle follows a pressure drop. The magma so created, rising up to the crust, feeds the volcanoes. In some types of margin between continents and oceans, the oceanic lithosphere is subducted, resulting in particular types of magma contaminated by the continental crust, called andesitic. Other volcanoes, such as those on Hawaii, are not on the margins, but in the middle of the plates (Fig. 3.29). These volcanoes are created by a warm plume at the border

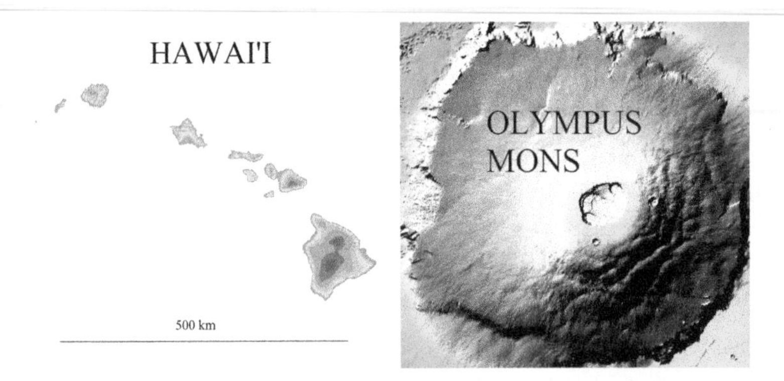

Fig. 3.29 The Hawaiian archipelago and Olympus Mons to the same scale. Left: image 120383081 FOTOLIA/pbardocz, scale and name added by author; right: MOLA (MSG), NASA

between the mantle and the outer core. The plume, which is warmer than the mantle surrounding it, rises by convection from the depths of the earth, a bit like smoke in the air. But since the plate moves over time over the fixed plume, the volcanoes are distributed along linear chains. The Martian volcanoes have a different distribution compared to Earth: not based on obvious plate lines (Mars does not have plate tectonics at present, and therefore the horizontal movements of the crust have been small), but rather on geographic areas.

On Earth, effusive volcanoes do not give rise to dangerous explosions, but rather to tranquil lava flows. The resulting rocks, which are dark and heavy due to iron-magnesium content, are known as basalts. Basaltic flows tend to run long from the vent before they cool down, so that they build a flat and very large volcanic building, known as a shield volcano. On Earth, one example of classic shields is precisely the Hawaiian ones. Most of the volcanic activity on Mars has been effusive, and many lava flows can be traced on the maps for hundreds of kilometers (Fig. 3.30). In a simple yet interesting experiment, one can build a desk model of a shield volcano with overlapping wax flows.[11] The analog model applies to any shield volcano (Fig. 3.31). In fact, perhaps it better suits Martian volcanoes without aureole, such as Alba Mons, while for Olympus Mons, the shape of the edifice was changed when the aureole and the basal scarp were created (The Fifth and Sixth Mysteries).

FIFTH MYSTERY: Explosive Volcanism and Other Volcanic Riddles

What determines whether a volcano will be explosive or effusive is mostly the silica content of the magma. Since free silica, chemically consisting of a silicon atom enclosed by four oxygen atoms, forms long chains (polymerization) that hinder magma flow, silica-rich magma is also the most viscous. Very siliceous magmas also have lower temperatures, which makes them even more viscous, as viscosity increases dramatically with decreasing temperature. Volcanoes with high silica

[11]Albin, E. F. (1998, March). Build Olympus Mons!. In Lunar and Planetary Science Conference (Vol. 29).

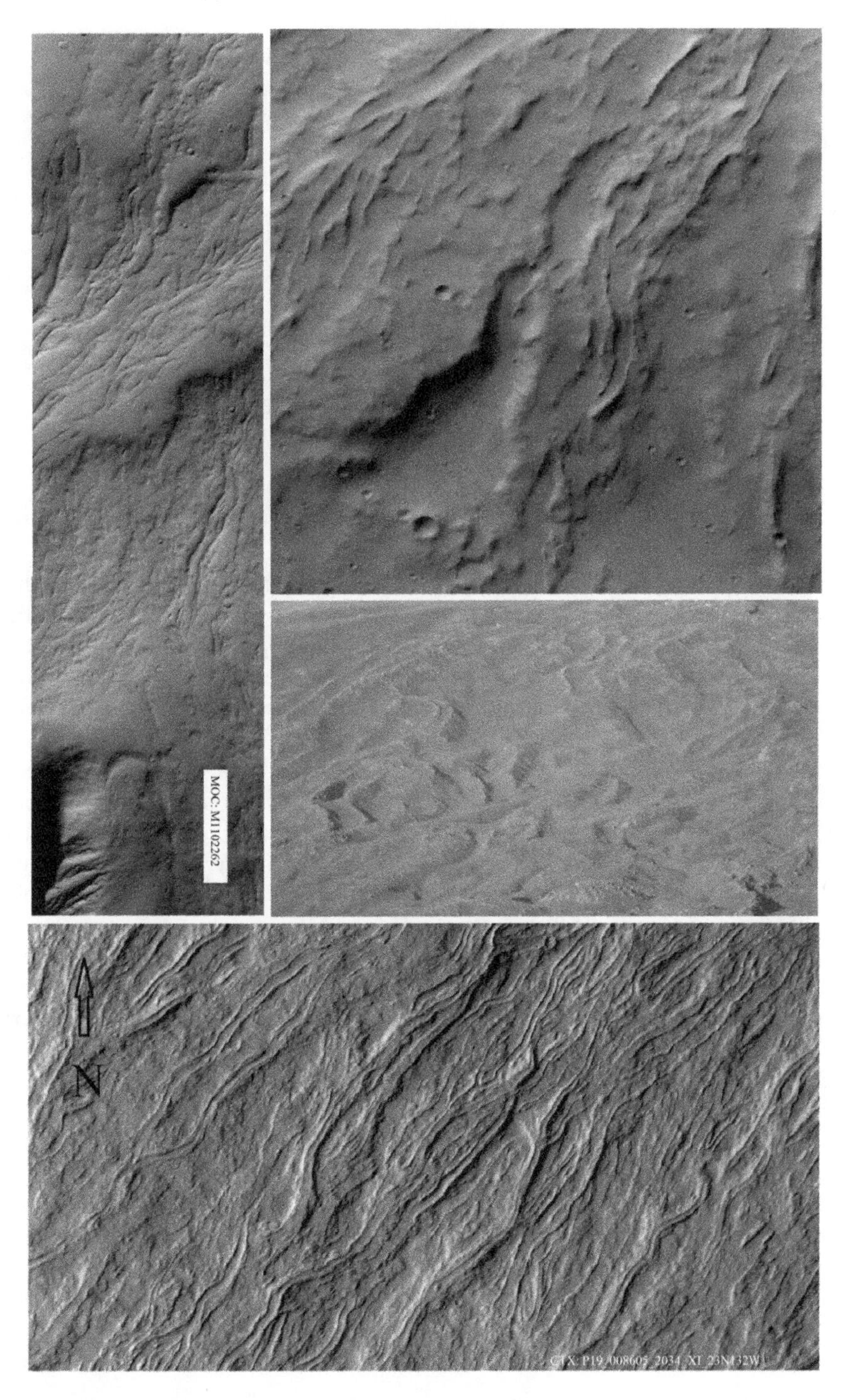

Fig. 3.30 Only one of these images is terrestrial. Lava flows can be seen on the north side of Olympus Mons in this MOC image (top left). Image width 3.25 km. Top right: detail of the previous image. Width of the image about 2 km. Bottom: other lava flows from Olympus Mons. Image width about 8 km. Middle: lava flows of Etna. MOC (Malin Space Science Systems, MGS), NASA, FVDB, CTX

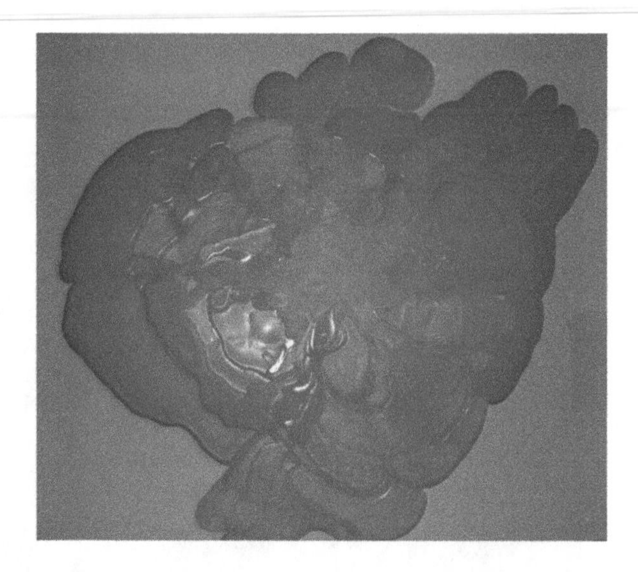

Fig. 3.31 Successive flows of hot wax provide an effective small-scale analog of a shield volcano. Exactly as in a real shield volcano, hot and fluid wax flows at the flank of the heap. However, it soon increases viscosity as it cools down with distance from the center, and comes to a halt at gentle slopes. Image about 70 cm across. FVDB

content respond to the internal pressures of the magma chamber very slowly, as the viscous magma behaves almost like a rigid plug. The pressure inside the magmatic chamber increases, until the plug is removed violently in a cloud of hot pyroclastic flows. Such particles deposit as tephra. Volcanoes whose magma is silica poor, on the other hand, produce the characteristic flows that can reach tens of kilometers from the eruptive mouth and little, if any, tephra. Most Martian volcanoes show effusive lava flows, but deposits from pyroclastic flows are also well represented. The volcano Ceraunius Tholus, for example (Fig. 3.32), shows erosion marks typical of water running along poorly consolidated tephra materials, similar to terrestrial volcanoes with similar composition (Figs. 3.33 and 3.34). Ceraunius and Uranius Tholi, shown in Fig. 3.33, have likely ages between 3.7 Ga and 4 Ga,[12] and this makes them among the most ancient volcanoes on Mars.

Differences between terrestrial and Martian eruption dynamics are poorly understood. On Earth, the difference between poor and rich silica volcanoes is explained in terms of magma contamination. Lack of silica occurs when the primary, silica-poor magma coming from the earth's mantle reaches the surface almost uncontaminated. Silica-rich magmas, in contrast, are contaminated by the continental, silica-rich crust, resulting in a more viscous and explosive magma.

But there is no trace of silica-rich rocks on Mars. How is the presence of explosive volcanoes possible? In the late 1990s, the data from Sojourner, the toy-sized rover from the Pathfinder mission, together with that elaborated by CRISM (Compact Reconnaissance Imaging Spectrometer for Mars), a chemical analysis tool on board the Mars Reconnaissance Orbiter, revealed the presence of andesites, rocks typical of the Andes, a mountain chain where the oceanic lithosphere is subducted beneath the South American lithosphere. How did these rocks form on Mars, if there is no plate tectonics? Another oddity is that it is hypothesized that the northern hemisphere was

[12]Werner, S. C. (2009). The global martian volcanic evolutionary history. *Icarus, 201*(1), 44–68.

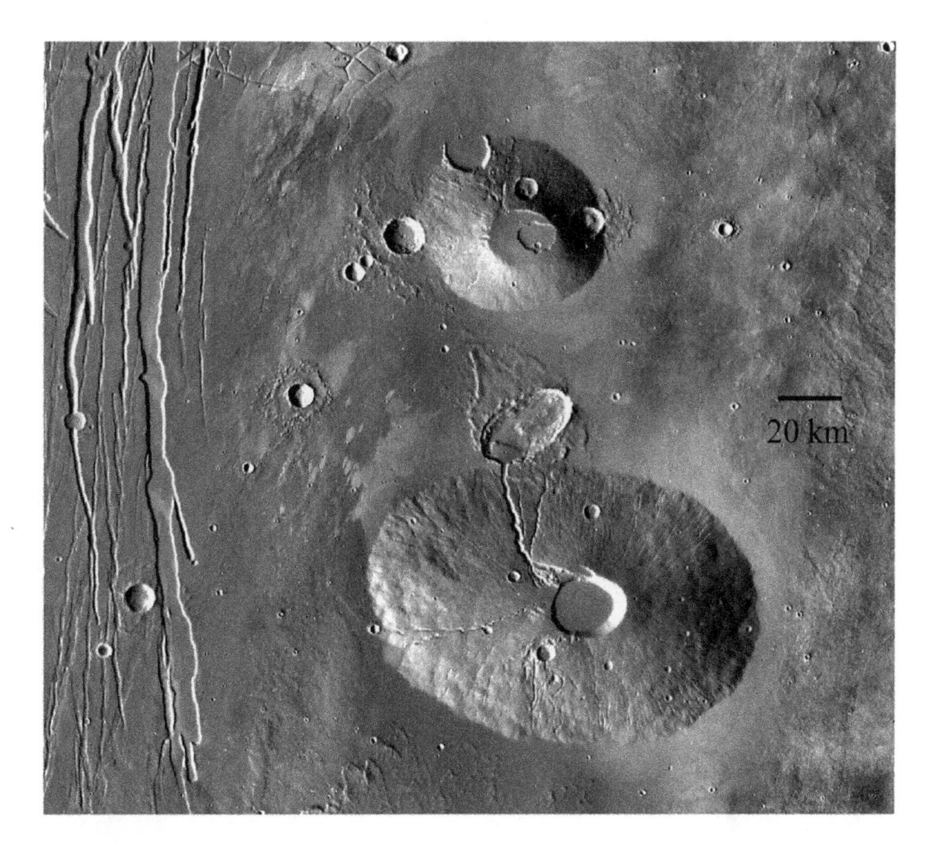

Fig. 3.32 This striking image shows the volcanoes Ceraunius Tholus at the bottom and Uranius Tholus at the top. Although they are not among the largest volcanoes on Mars, they are huge by terrestrial standards. Ceraunius is as wide as one fourth of Switzerland. Note the radial gullies on Ceraunius Tholus. A meandering torrent to the north has deposited a fluvial delta inside an impact crater, which by itself belongs to the relatively rare class of oblong craters with layered ejecta! Thus, we can see in the same image at least five remarkable morphologies. The gully is shown in better detail in Fig. 3.28. CTX, NASA

the bottom of an ancient ocean (Chap. 4). On Earth, oceanic rocks are basaltic, and not andesitic. Thus, there would be a strange reversal: the rocks typical of the Earth's oceans would, on Mars, emerge on the continents; and the continental land rocks would be at the bottom of the ancient ocean. Another explanation is that CRISM did not, in fact, detect andesites, but rather basalt altered by phyllosilicates, perhaps as a simple result of the Martian ocean water. In fact, the determination of mineralogical composition with remote sensing is not univocal. An appropriate combination of clay minerals resulting from aqueous alteration of basalts could simulate andesites.

However, there is indeed tephra on Mars. How to explain its presence alongside poorly siliceous magmas? This is not well understood, but we know at least one reason why volcanoes on Mars could produce tephra more easily than on Earth. The gas content in a magma is one of the driving forces of an eruption. At a certain depth, magmas contain dissolved water and carbon dioxide, loosely referred to as volatiles.

Fig. 3.33 Pyroclastic material is easily erodible by running water. The Bromo, Batok, and Semeru volcanoes on Java, Indonesia. Compare with the erosion marks in Fig. 3.32. Image 116691110 FOTOLIA/SANCHAI

Fig. 3.34 Tephra is poorly consolidated. Quarry in central Italy. FVDB

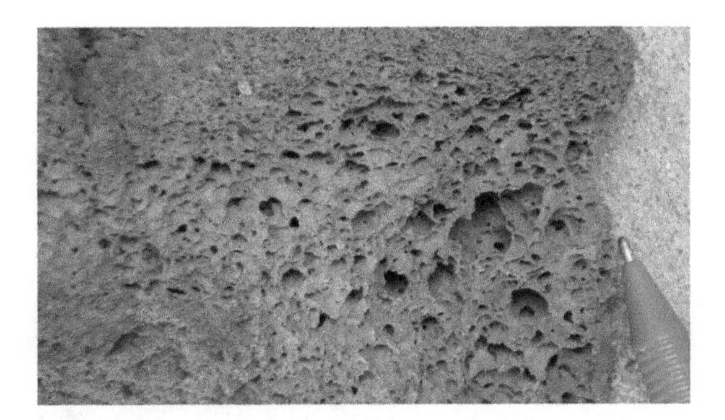

Fig. 3.35 Bubbles in a magma. Etna, Sicily. FVDB

When the magma comes closer to the surface, the pressure drops, meaning the volatiles in the magma are no longer stable in liquid form, and thus become a gas. The magma begins to bubble, forming a froth (Fig. 3.35). This process drives the final stage of the eruption, with the bubbles floating to the top, and so thrusting the magma violently toward the surface. A proper analogy is with a soda bottle opened on a summer day. Because of the reduced gravity field, on Mars, the nucleation of bubbles occurs more easily and earlier during the eruption compared to Earth. Thus, tephra probably result from a basaltic eruption on Mars, when the same magma composition on Earth would have created a tranquil lava flow.

We come now to another possible recent enigmatic finding, which could expand our view of Martian volcanology. Supervolcanoes are a kind of volcano that gives off eruptions for which we, fortunately, have no direct experience. The infamous, extremely violent eruption of the Indonesian volcano Tambora in April 1815, probably the largest in the Holocene, was so huge in terms of the tephra released into the atmosphere that airborne particles were responsible for a year without a summer, and yet it was much milder than a supervolcano. We can only study the volcanic edifices of such super volcanoes, which are endowed with huge calderas. Supervolcanoes are capable of producing thousands of cubic kilometers of tephra in just one single eruption. They are so violently explosive that the material is normally ejected all the way up into the stratosphere, with important consequences for the climate and at times even for species survival. In fact, supervolcanoes may be responsible for significant species extinctions among plants and animals. According to a theory, one eruption of the Toba supervolcano in Sumatra about 84,000 thousand years ago may have reduced human diversity, bringing our species to the extinction threshold (Fig. 3.36). Another supervolcano lurks beneath Yellowstone National Park.

Recently, it has been suggested[13] that some craters in Arabia Terra, so far loosely interpreted as impact craters (or, more accurately, with no skepticism having been expressed as to their being impact craters), could be volcanic. The largest of such

[13]Michalski, J. R., & Bleacher, J. E. (2013). Supervolcanoes within an ancient volcanic province in Arabia Terra, Mars. *Nature*, *502*(7469), 47–52.

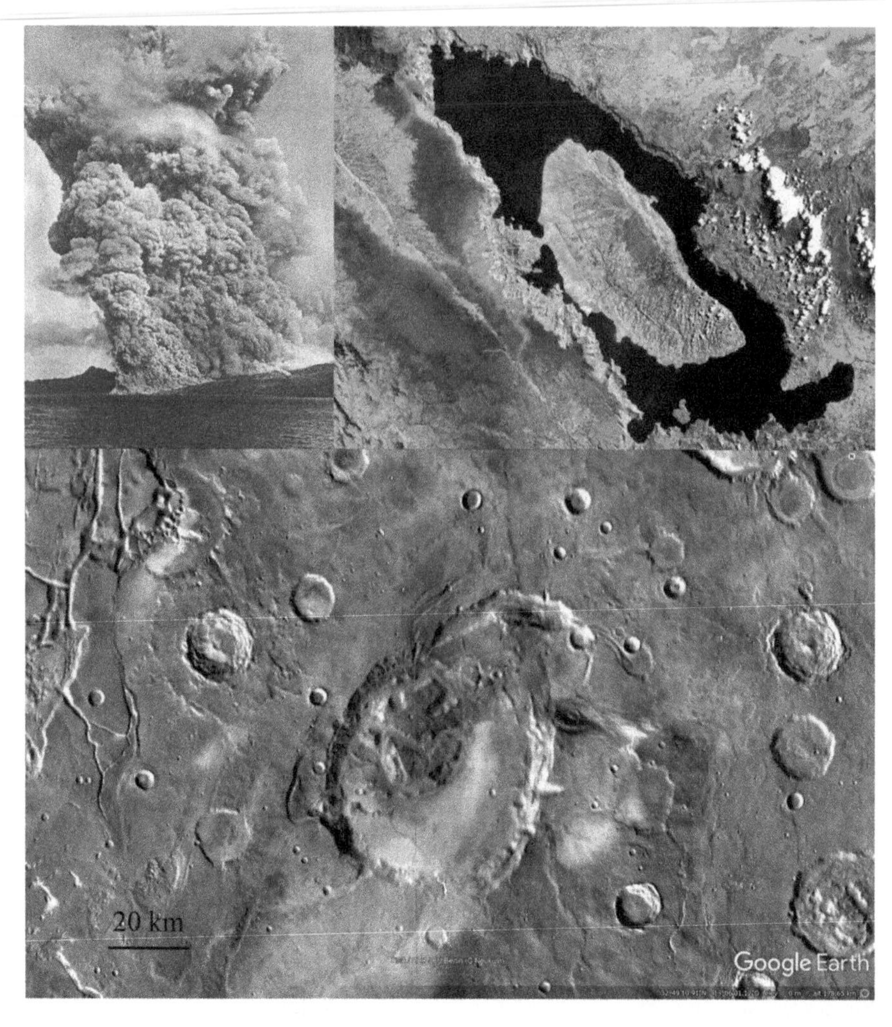

Fig. 3.36 Top left: pyroclastic eruption from Mount Pelee. Top right: the Toba caldera, Sumatra. Bottom: Eden Patera, CTX mosaic on Google Mars platform. Top left: Delacroix, public domain; top right: LANDSAT, NASA; bottom: CTX (MRO, NASA), composite image on Google Mars platform

craters is the irregular, very large (50×80 km) Eden Patera (Fig. 3.36). The bizarre shape, the presence of concentric fractures more typical of gravitational flank instability than impact, the peculiar depth-to-width ratio (an impact crater of this size should be deeper), and geochemical signatures indicate that this crater could be volcanic rather than meteoritic. Considering its similarity with the irregular calderas such as Toba, could Eden Patera and other craters in Arabia Terra be Martian supervolcanoes? Supervolcanoes and their associated super-eruptions may have shed a great deal of tephra in Arabia Terra. This could solve some of the enigmatic features of the area, such as the presence of fretted terrain and of layered sediments

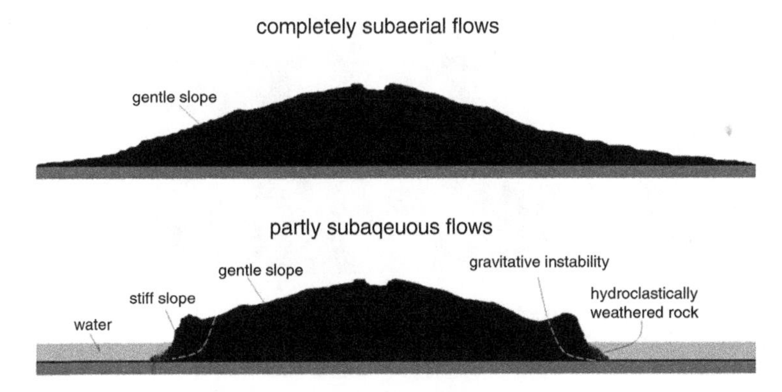

Fig. 3.37 The present basal scarp surrounding Olympus Mons is the scar of the aureole landslide deposits. However, while a completely subaerial flow forms a gently-sloping volcanic edifice like the one of Alba Mons (top), the presence of a medium like water may have altered the otherwise gentle slopes and also weathered the basalt, causing the gravitational instability. FVDB

bearing sulphates and clay. These sediments could be the deposits of massive super-eruptions. Why would there be supervolcanoes here? Could it be related to the thinner crust associated with Arabia Terra, which is somehow a transitional region between the southern highlands and the northern plains?

A further volcanic mystery regards the Olympus Mons volcano. We have seen that Olympus Mons is a shield volcano created by a stacked sequence of low-viscosity lava flows. The structure of Olympus Mons reveals a unique feature of the Martian volcanoes. While the central part of the volcanic edifice slopes with angles gentler than 1^0–5^0 degrees, it becomes much steeper at the periphery of the edifice, where a basal scarp appears. Moreover, Olympus Mons is the only volcano bordered by an aureole. The aureoles have been interpreted as landslide deposits (Sixth Mystery) and the basal scarp as the scar of landslides. However, there is a riddle here.

A volcanic shield like the one shown in Fig. 3.37 featuring a "gentle slope," as indicated in the figure (typically at 1°–5°), is stable, i.e., it would not fail gravitationally. However, if aureoles are the deposits of gigantic landslides, failures did occur, producing the largest landslide deposits in the solar system! What created the necessary instability in the first place? The best terrestrial analogs are the Hawaiian islands. Similar to Olympus Mons, Hawaii is surrounded by an aureole of landslides too, albeit much smaller ones. The difference between the aureole-bearing Hawaii and other basaltic volcanoes devoid of any aureoles is the oceanic nature of Hawaii, namely the presence of water. When a basaltic lava flow ends up in water, it cools much more rapidly than in air. Solid lava is dumped into the vicinity of the shore, where it accumulates and piles up together with previous flows. Moreover, this fast-solidifying lava forms a kind of weak rock called hyaloclastite. For these two reasons, this rock heap is, therefore, unstable and fails, producing a major landslide, an aureole (Fig. 3.37). The base of Olympus Mons falls in the area proposed for the Oceanus Borealis. Can this be another clue as to the presence of an ocean on Mars in the early Amazonian, the time of formation of the aureoles? It is, however, not

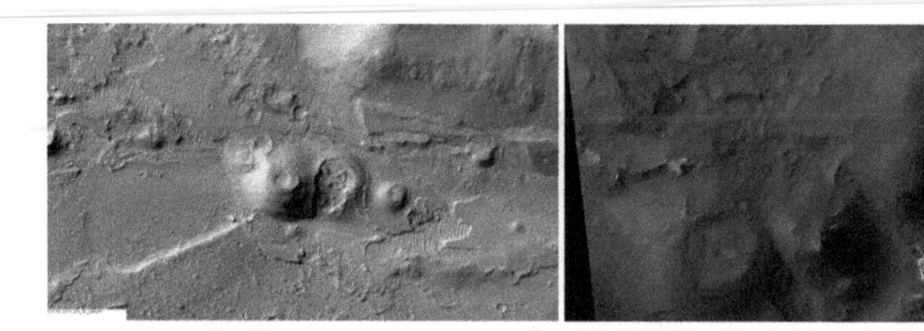

Fig. 3.38 Strange cones in Coprates Chasma. Left: CTX image. Right: HiRISE. Left: CTX (MRO, NASA), right: HiRISE (MRO, NASA)

precisely clear as to how this may have occurred, since the ancient ocean was perhaps only 1 km deep in that position, while the basal scarp is about 8 km high.

There are many other volcanic mysteries on Mars. One recently found enigmatic feature is visible in great detail on HiRISE images, and can also be seen with lower resolution in CTX photos (Fig. 3.38). These features are volcanic-like cones in Coprates Chasma, inside Valles Marineris. The presence of volcanic cones would open up many questions, because even though Valles Marineris occupies the middle of the Tharsis bulge, there are no indications of volcanoes in Valles Marineris. Moreover, such cones are small and require a small magma reservoir. Thus, was Valles Marineris by chance carved just above a small volcanic reservoir? Nevertheless, these cones in Coprates can also be interpreted in a non-volcanic manner. Could they be pingos, i.e., morphologies due to ice injections, rather than volcanoes?

As early as the end of the sixteen century, the philosopher Giordano Bruno, pointing out that most volcanoes known at that time occurred in proximity to the coastline, invoked a natural explanation. Three and a half centuries later, plate tectonics explained that part of the volcanoes occur at the margin of active plates. Thus, the distribution of terrestrial volcanoes has aided in understanding the very nature of volcanoes as a geological phenomenon.

Yet, there is another mystery. In contrast with terrestrial volcanoes, whose distribution is understood in the frame of global tectonics, the distribution of the volcanoes on Mars remains enigmatic. It is a fact that the largest volcanic province in the whole solar system, Tharsis, is approximately antipodal with respect to one of the largest impact basins of the solar system: Hellas Planitia. The correspondence is nearly exact with Alba Mons, one of the most representative of Tharsis' volcanoes. A less precise but still intriguing antipodality relationship occurs between the other huge impact basin on Mars, Argyre Planitia, and the volcanic region of Elysium. Might the extraordinary impact of the asteroid that created Hellas Planitia (an energy on the order 10^{26} to 10^{27} joules) have generated shock and seismic waves that travelled to the opposite side of the planet, where they may have melted magma as a consequence of the pressure variation carried by the waves? Other fascinating correspondences arise from our own planet. The impact crater of Chixculub is approximately the same age as the so-called Deccan traps. These are enormous expansions of basaltic lava that inundated the Deccan region of India in the late Cretaceous. Both the impact that created Chicxculub and the traps have been separately invoked to explain the mass extinction at the end of the Cretaceous, including the disappearance of the dinosaurs. The two locations were very approximately antipodal when the impact occurred. The latitudes of the impact site and of the traps were at the end of the Cretaceous +30° and −30°, respectively (perfect correspondence), while longitudes were about 50° off (130° apart in place of 180°), which may be too much. However, it should be stressed that these reconstructed positions are affected by uncertainties. The distributions of volcanoes on Mars remains a mystery.

SIXTH MYSTERY: Enigmatic Mountains and Terrae

Figure 3.39 better shows the rugged halo surrounding the Olympus Mons volcano. It has been called the aureole in relation to Latin Aurum ("Gold"). From Aurum, we derive the adjective Aureus ("Golden"), and then "aureole" to indicate the mystic cloud around the images of sacred personages. It encircles the Solar system's highest and most enigmatic volcano. A look at the aureole shows that it is a composite deposit made up of lobe-shaped subunits separated by neat boundaries. The largest Western unit (W in Fig. 3.36) stretches for nearly 700 km from the border of the volcano. It has been successively blanketed by other smaller units, notably the N and the NW. Note also the presence of nearly completely buried aureoles to the east and

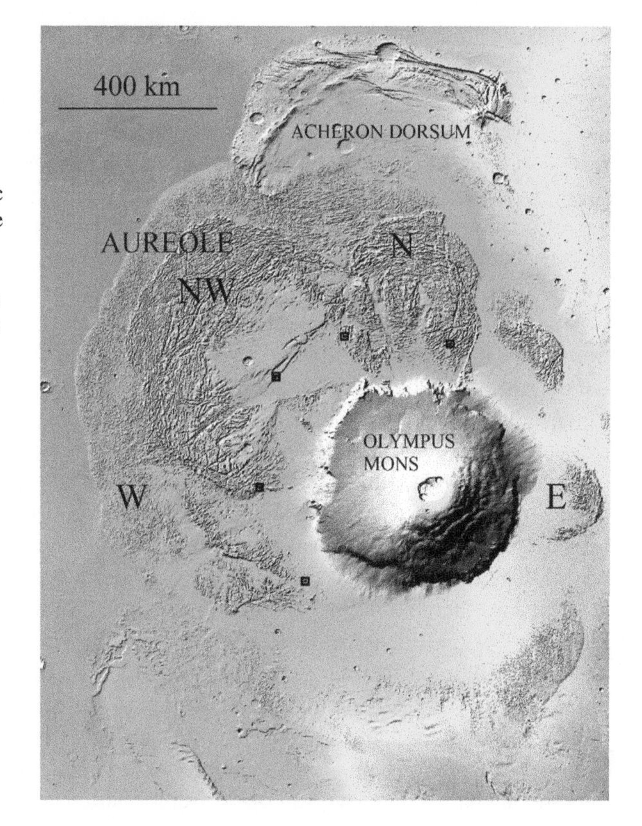

Fig. 3.39 Olympus Mons, with its aureoles, some of which have been labeled (see text). The squares represent the length of the volcano if the aureoles are restored back to the volcanic edifice, and may indicate the position of the Olympus Mons edges before the catastrophic landslides. The superposition of some units shows that some aureoles were created first, while others belong to a later stage. MOLA (MGS, NASA), mdf

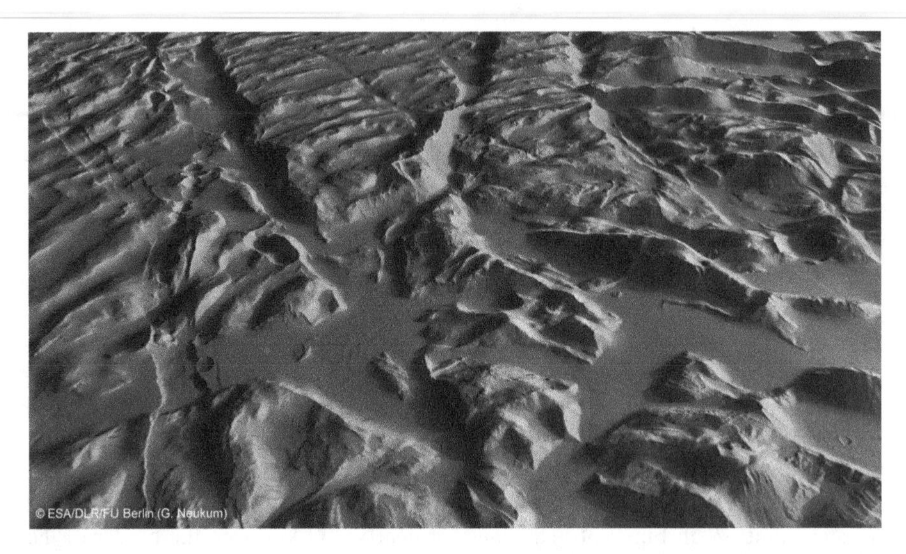

Fig. 3.40 The aureole appears as a rugged deposit with parallel ridges and partially buried by Aeolic deposits. HRSC image aboard Mars Express of the E aureole. HRSC (Mars Express, ESA/DLR/FU)

south under thick lava flows from Olympus, Pavonis, and Arsia Montes (the eastern E aureole is shown). Considering that the aureole is about 1 km thick (a thickness of more than 2 km is reached at some locations), and that the extension is slightly less than one million square kilometers or nearly one tenth the extension of Europe, the total resulting volume is between one half million to one million cubic kilometers (semi-buried aureoles certainly may contribute significantly). A close look at the aureoles shows high transverse ridges characterizing the "ruggedness," shown in greater detail in Fig. 3.40.

The aureole is markedly asymmetric: the northward and westward aureole lobes are long and wide, whereas the southward and eastward units are smaller. This asymmetry follows the local gradient, approximately directed from SE to NW. The low thermal inertia, widespread dark streaks, and the CRISM identification of olivine and pyroxene suggest that the aureole's surface is composed of fine-grained regolith derived from the weathering of basalt. What is the origin of these bizarre deposits?

One initial explanation is volcanic. We have seen that Olympus Mons has ejected very long lava flows along different directions. Could the aureole possibly be the result of giant lava flow deposits? Lava flows, both on Earth and on Mars, are smooth on the length scale of several kilometers. Blocks decameters in length may be identified on the aureole, but these are much smaller than the ruggedness of some kilometers of the ridges, so characteristic of the aureole deposits. However, it has been suggested that the ruggedness could perhaps be the result of eruptions taking place under particular conditions, for example, under ice. Other volcanic origins proposed for the aureoles are deposits of pyroclastic flows.

Other explanations are based on gravity rather than volcanism. Some researchers interpret the aureole as the result of slow deformation and lateral spreading, perhaps induced by the extreme weight of Olympus Mons. Other authors have noted the similarity of the aureole to landslides deposits, whereas the basal scarp (another enigmatic feature "Fifth Mystery") might be the scar of such landslides. This is probably the most accepted view at present. At first, it is difficult to accept the idea that an 8–9 km-high failure may have travelled for several 100 km (the length of France measured from north to south) along a flat area. A failure of this geometry would correspond to a friction coefficient on the order 0.012, fifty times smaller than the one of rock, when in principle, they should be the same! This may appear to be a fatal assessment for the landslide model. However, large landslides are also known to travel with reduced friction on Earth (Sect. 3.3). As we saw, this poorly explained feature is also a mystery for terrestrial landslides, even though in the terrestrial case, the reduction from the friction coefficient of bare rock to that of large rock avalanches is, at most, a factor of ten. Still a lot, but not as extreme as for the aureoles, which need five times more reduction.

One possibility is that the aureoles are subaqueous and not subaerial landslides. Subaqueous landslides are known to travel even longer due to the lubrication effect of water. There are numerous ideas as to how this could take place, but the details of the process are still unclear. The consequences could be interesting, because it would imply that the Oceanus Borealis was there during the gigantic failures that produced the aureole landslides. However, according to the proponents of the Oceanus Borealis, some of the proposed shores for the Oceanus during the Noachian extended to Amazonis Planitia and the surrounding areas, where the aureole landslides subsequently took place. The problem is that the aureoles are dated at a much later time, perhaps as close to us as 3.5 or 3 billion years, or perhaps even younger. Early measurements based on Viking showed only 51 craters with diameters wider than 1 km for the whole aureole, which, considering the huge area, would give an age of only a few hundred million years! Nevertheless, photographs such as the one in Fig. 3.40 show that the aureoles are filled with Aeolic deposits, which have certainly masked many, perhaps most, of the craters that would be overlooked. Notice that the aureole deposits are unique on Mars. The other Martian volcanoes, such as Alba Mons, have no such aureoles. Perhaps the aureole is similar, albeit at a much greater scale, to the landslide deposits surrounding volcanic islands in the terrestrial oceans, the Hawaii and Canary Islands.[14]

[14]For the aureoles, see Hiller, K. H., Janle, P., Neukum, G. P. O., Guest, J. E., & Lopes, R. (1982). Mars: Stratigraphy and gravimetry of Olympus Mons and its aureole. *Journal of Geophysical Research: Solid Earth*, *87*(B12), 9905–9915.; Mouginis-Mark, P. (1993, March). The influence of oceans on Martian volcanism. In *Lunar and Planetary Science Conference*(Vol. 24).; McGovern, P. J., Smith, J. R., Morgan, J. K., & Bulmer, M. H. (2004). Olympus Mons aureole deposits: New evidence for a flank failure origin. *Journal of Geophysical Research: Planets*, *109*(E8).; De Blasio, F. V. (2011). The aureole of Olympus Mons (Mars) as the compound deposit of submarine landslides. *Earth and Planetary Science Letters*, *312*(1–2), 126–139; Mouginis-Mark, P. J. (2017). Olympus Mons volcano, Mars: A photogeologic view and new insights. Chemie der Erde, https://doi.org/10.1016/j.chemer.2017.11.006, 35 pp

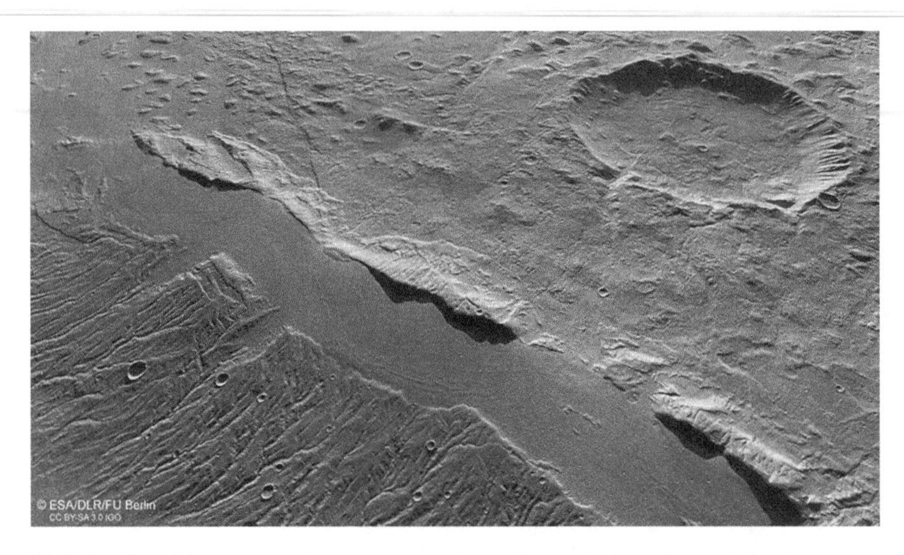

Fig. 3.41 One of the many trenches in western Acheron Fossae that have dissected the ridge. Inside the trench, there is evidence of plastic flow. Note the numerous craters and the erosion marks downslope of the trench, apparently in poorly consolidated material. Image Mars Express 367990, courtesy of ESA/DLR/FU Berlin, CC BY-SA 3.0 IGO. Scene about 30 km across. HRSC (Mars Express, ESA/DLR/FU)

A final remark. If the aureoles are the result of landslides from Olympus Mons, the question arises as to the extension of the volcano prior to failures, which can be estimated if the rock in the aureoles is restored back to the volcano. Fig. 3.36 shows the pristine boundary of the volcano reconstructed in this way. Probably, Olympus Mons was originally more than 200 km wider, totaling a width at its maximum expansion of more than 800 km[15]!

Not far from the aureole lies another enigmatic feature, called Acheron Fossae, in reference to the series of trenches crossing it[16] (Figs. 3.39 and 3.41). Acheron is a ridge (also called "Acheron Dorsum" by some authors) standing about 2–3 km from the plains of Arcadia Planitia to the north and west, while its eastern and southern boundaries are in contact, respectively, with lavas from Alba Mons and the aureole seen earlier. On Earth, we would simply call it a "mountain." A very extensive mountain, indeed: both its length of 800 and width of 350 km are remarkably comparable to the dimensions of the Alpine mountain range. Similar to the Alps, Acheron Dorsum has a bowed shape with a radius of curvature of about 600–800 km. However, in contrast to the Alps, whose formation is understood

[15]De Blasio, F. V. (2018). The pristine shape of Olympus Mons on Mars and the subaqueous origin of its aureole deposits. *Icarus*, *302*, 44–61.

[16]Kronberg, P., Hauber, E, Grott, M., Werner, S.C., Schafer, T., Gwinner, K., Giese, B., Masson, P., Neukum, G., 2007. Acheron Fossae, Mars: Tectonic rifting, volcanism, and implications for lithospheric thickness. Journal of Geophysical Research 112, E04005, doi:10.1029/2006JE002780, 2007

within the framework of plate tectonics, Acheron Dorsum appears somehow enigmatic.

Acheron is ancient. Its antiquity can be appreciated by the high number of impact craters, especially in comparison to the surrounding plains of Arcadia and Amazonis Planitiae. On Acheron, there are also river networks. An age of 3.4–4 Gyr results from crater counting, corresponding to the late Noachian-Hesperian.[17] Thus, Acheron is probably older than the aureole deposits, with an estimated age of 1.5–3.6 Ga.[18] Acheron is split by numerous fossae, wide and deep trenches that cross the whole ridge along different directions, with a prevalent east-west direction. Such trenches are peculiar, because they are apparently unrelated to the fractures outside Acheron. They may be either the response to currents in the mantle underneath the ridge or the product of gravitational re-adjustment of the ridge, similar to a slowly opening landslide. If the origin of the trenches is poorly understood, there is no obvious explanation for the origin of Acheron itself, either. The ridge is in the midst of the volcanic region of Tharsis, yet there seems to be no direct relationship with the other volcanoes. Because of its arcuate shape, Acheron has been envisaged as the expression of the stress created by Olympus Mons. One of the possible problems is the age difference between Olympus Mons, regarded as recent (although the core could be much older!), and the much older Acheron. Another theory is that Acheron is the remnant of an ancient caldera. We know that calderas are the expression of explosive volcanism. On the one hand, explosive caldera-producing volcanoes are not the typical volcanic style on Mars; on the other hand, explosive volcanoes may be present in Arabia Terra (Fifth Mystery), even though the suspected Eden Patera caldera (85 km across) is much smaller than Acheron Fossae. Another hypothesis is that Acheron is the remnant of an ancient landslide from Olympus Mons, a sort of pre-aureole failure. This would explain the arcuate ridge and the apparent fragmented state of most of Acheron, but would imply a very ancient (Noachian) Olympus Mons, and require the landslide to have traveled a very long distance.[19] The Acheron ridge is an example of a feature on Mars that, while not among the largest (although it isn't small, either), may reveal much about the history of Mars once we shed more light on its enigmatic features.

[17]Hiller, K. H., Janle, P, Neukum, G.P.O., Guest, J.E., Lopes, R.M.C., 1982. Mars: Stratigraphy and gravimetry of Olympus Mons and its aureole. JGR 87, 9905–9915; Acheron corresponds to the unit lNh (late Noachian Unit) in the USGS geological Map of Mars. of Tanaka et al., 2014.

[18]Morris E.C., Tanaka, K.L., 1994. Geologic maps of the Olympus Mons region of Mars: U.S. Geological Survey Miscellaneous Investigations Map I-2327, scale 1:2,000,000

[19]For recent discussions, see McGovern, P.J., Smith, J.R., Morgan, J.K., Bulmer, M.H., 2004. Olympus Mons aureole deposits: New evidence for a flank failure origin. J. Geoph. Res. 109, E08008; Isherwood, R.J., Jozwiak, L.M., Jansen, J.C., Andrews-Hanna, J.C., 2013. The volcanic history of Olympus Mons from paleo-topography and flexural modeling. Earth and Planetary Science Letters 363, 88–96; Scott et al. (1981) and Howard (1981), De Blasio, F. V., & Martino, S. (2017). The Acheron Dorsum on Mars: A novel interpretation of its linear depressions and a model for its evolution. Earth and Planetary Science Letters, 465, 92–102.

Addenda

- The bottom of the huge impact basin called Hellas Planitia is the largest of its kind in the whole solar system. It is filled with sediments, dust, and regolith, to the point that on the bottom, there is often a powder mist suspended by the Martian winds.
- Since the formation of Valles Marineris, there has apparently been no horizontal continental movement similar to the terrestrial continental drift.
- Landslides in Valles Marineris are so huge that the comminuted rock enclosed in just one of the largest landslide deposits might possibly be able to blanket the whole European surface with a layer of 10 cm.
- Because most Martian volcanoes have had innumerable eruptions, the age of their formation is uncertain, as the products of the earliest eruptions are completely covered by latest ones.
- Olympus Mons is so broad that, walking on its surface, we would not realize that we were on a shield volcano, as most of the volcanic edifice would be hidden beyond the horizon.
- Volcanoes where water and heat have co-occurred may be a good first place to look for past and present life on Mars.

Chapter 4
Ice, Water

On June 15 and 19, 2008, the Phoenix mission lander dug a small trench in the hard ground of the Martian Arctic highlands. Immediately, a white patina became visible. After four sols, an enlarged image showed that part of the whitish veneer had disappeared. Only water ice can show this sort of behavior. The patina was transformed directly from a solid to a gaseous state, skipping the intermediate fluid state, a phenomenon that occurs at low pressure and that is known as sublimation. The experience proves that at those high latitudes, the Martian soil is impregnated with ice. The search for water and ice on Mars is one of the main obsessions in regard to the Red Planet, and the reason is simple: as far as we know, life requires water. The largest reservoir of water may have been an ancient ocean that covered the northern lowlands of Mars for over 25% of the planet's surface. How do we know of the past existence of this ocean? When and why did a huge amount of water disappear, leaving the barren and dry land we see today?

On the pages that follow, we will see that not only have large amounts of ice been found on ground of high latitude on Mars, but that ice has also been common at lower latitudes. Strange glacial tongues similar to terrestrial rock glaciers, degraded craters due to ice relaxation in their core, enigmatic chaotic zones: they all seem to show the importance of ice in the evolution of the Martian surface. But there is also evidence of periods when water flowed in the liquid state: torrents similar to earthly drainage networks, long channels hundreds of kilometers long excavated by violent streams, and temporary patterns formed by local ice melting. Perhaps there were lakes on Mars, and most probably an entire ocean. How did the water disappear?

F. V. De Blasio, *Mysteries of Mars*, Springer Praxis Books,
https://doi.org/10.1007/978-3-319-74784-2_4

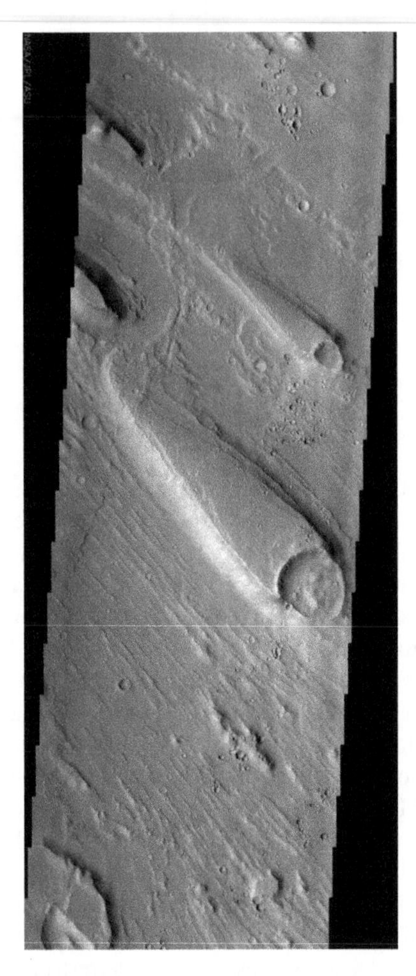

The image shows the tear-drop erosion that forms around craters, in this case, due to a water current that was travelling from bottom right to top left. THEMIS image. THEMIS (Mars Odyssey), NASA, NASA/JPL-Caltech/Arizona State University

4.1 Ice on Mars

Arctic Mars

Figure 4.1 shows a well-known picture of the Arctic area of Mars Scandia Colles, part of the enormous northern plains known as Vastitas Borealis. It displays a hole dug by the Phoenix mission in 2007. The difference between the two photographs taken at sol 20 and 24 shows that under the Martian soil, the ice is stable, but once it is brought to the surface, at low temperatures, it sublimes (i.e., the ice becomes vapor in the atmosphere without melting) within a few sols of time. On Mars, sublimation

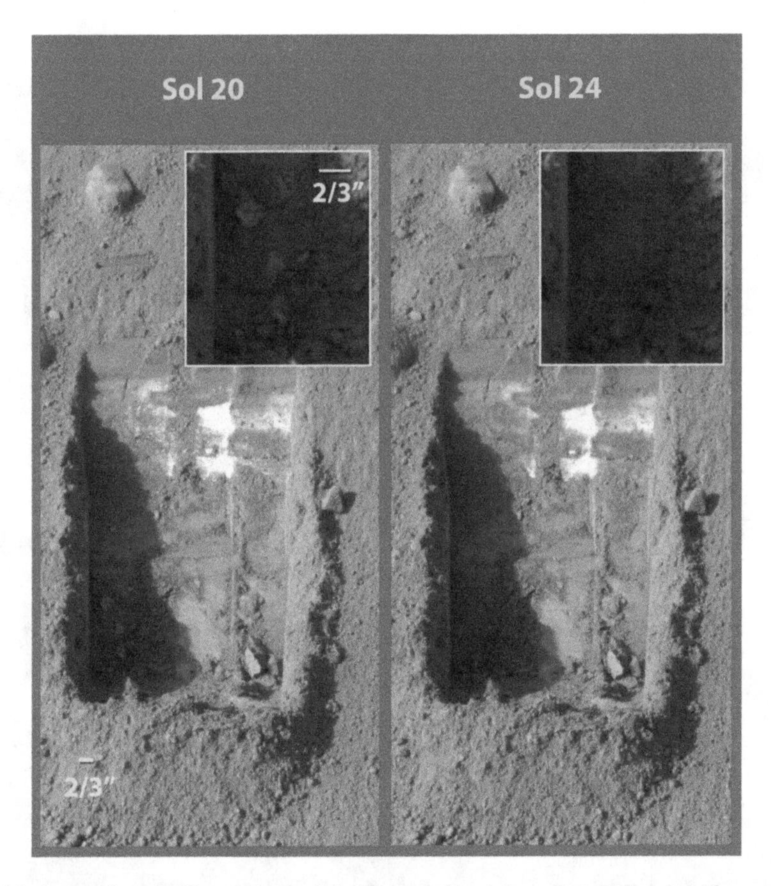

Fig. 4.1 Images from the Phoenix lander (mission headed by Arizona State University) taken on June 15 and 19, 2008 (sol 20 and 24 of the mission). The lander dug a hole (nicknamed Dodo-Goldilocks) under the arctic surface, revealing the presence of a white patina. After 4 days, an enlarged image showed that a part of the patina (the one below in the figure) had disappeared. Phoenix's images show that at high latitudes, Mars' soil is impregnated with ice. The reason why ice disappears within a few days is explained by the phenomenon of sublimation. Courtesy of NASA/JPL-Caltech/University of Arizona/Texas A & M University. Phoenix (NASA; Arizona State University)

is expected, because the Martian atmosphere is so poor in water that if the entire amount of water in the atmosphere were spread on the surface, it would create a layer only one hundredth of mm thin. What was less expected, at least in such evident terms, was the vivid image of ice impregnating the soil. Phoenix's direct observation leaves little doubt, but it cannot, of course, be conducted systematically, so other methods have to be used to determine whether or not ice is present on the Martian surface.

Fortunately, remote sensing data and the examination of morphology makes it possible to infer whether ice is present under the surface in different locations. Figure 4.2 shows a comparison between the plains of Vastitas Borealis photographed by Phoenix and the surface of the Arctic tundra. Both show the presence of patterned ground, or broken polygon soil, a typical morphology of permafrost. The next figure, 4.3, also shows a magnification of one of the polygons seen from Phoenix.

Fig. 4.2 The top figure shows the patterned ground imaged by Phoenix. The image has been exposed at different times to show the midnight sun on Mars (the fact that these areas are named Scandia Colles is particularly evocative). The bottom figure shows an example of patterned ground in the Arctic tundra. Top: Phoenix (NASA; Arizona State University), bottom: image 141637438 FOTOLIA/drepicter

Polar Caps

The ice caps of Mars were known to Huygens, who drew them in 1659 (drawing reported at the beginning of the first part). They are among the details best represented by observers of the nineteenth and twentieth centuries, who had well documented the greater amplitude of the north compared to the south cap (Fig. 4.4, top). Today's measurements provide more precise data: over a thousand kilometers in diameter and four hundred for the north and south cap, respectively (Fig. 4.4, bottom). This difference is not surprising if we consider that at perihelion, Mars turns its south pole towards the sun. The higher the radiation received, the greater the melting of the ice, the smaller the size of the cap.

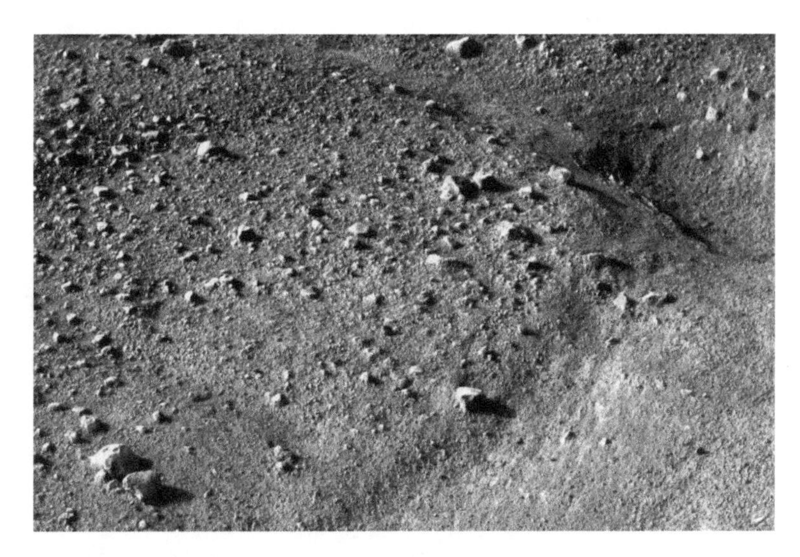

Fig. 4.3 Detail of one of the polygons imaged by Phoenix. Phoenix (NASA; Arizona State University)

Ice caps are important for understanding the atmospheric movements and, in the long term, the Martian climate. However, many details have only recently come to be understood and some issues are still unresolved. The caps are not only made of ice water, but are also covered by a dry ice CO_2 coating. This is the atmosphere of Mars itself, composed of the same carbon dioxide CO_2 gas that seasonally condenses in the cold season and then sublimes as temperatures increase.

The northern cap has a thickness of almost 3 km and is stratified. The ice has settled for millions of years, and in doing so, it has trapped airborne sediments. They may be a real resource for future explorers searching not only for water, but also for knowledge about the climatic history of the planet. In fact, thanks to the radar analysis of the SHARAD radar aboard the MRO mission, it was found that the top 300 m layer of polar caps is made up of ice deposited under "normal" conditions, while the underlying ice exhibits a different deposition regime. The level at 300 m therefore marks a shift to a different glacial regime, like the glacial–interglacial earth transition. This should not be surprising, considering that the inclination of the Martian rotation axis changes chaotically (Chap. 5), and therefore large climate variations are expected, superior to those here on Earth.

The extraordinary spirals on the north cap (Fig. 4.5) are carved by the action of catabatic winds created by differences in the temperature of the atmosphere. Leaving the Planum Boreum (i.e., the northern ice cap) at high speed, the winds are diverted to the west by the Coriolis acceleration, and in doing so, they erode the ice. The southern cap is eccentric by 150 km from the real pole of the planet. This strange location is probably due to the presence of the vast basin of Hellas Planitia, capable of diverting the masses of air and favoring its skewed deposition.

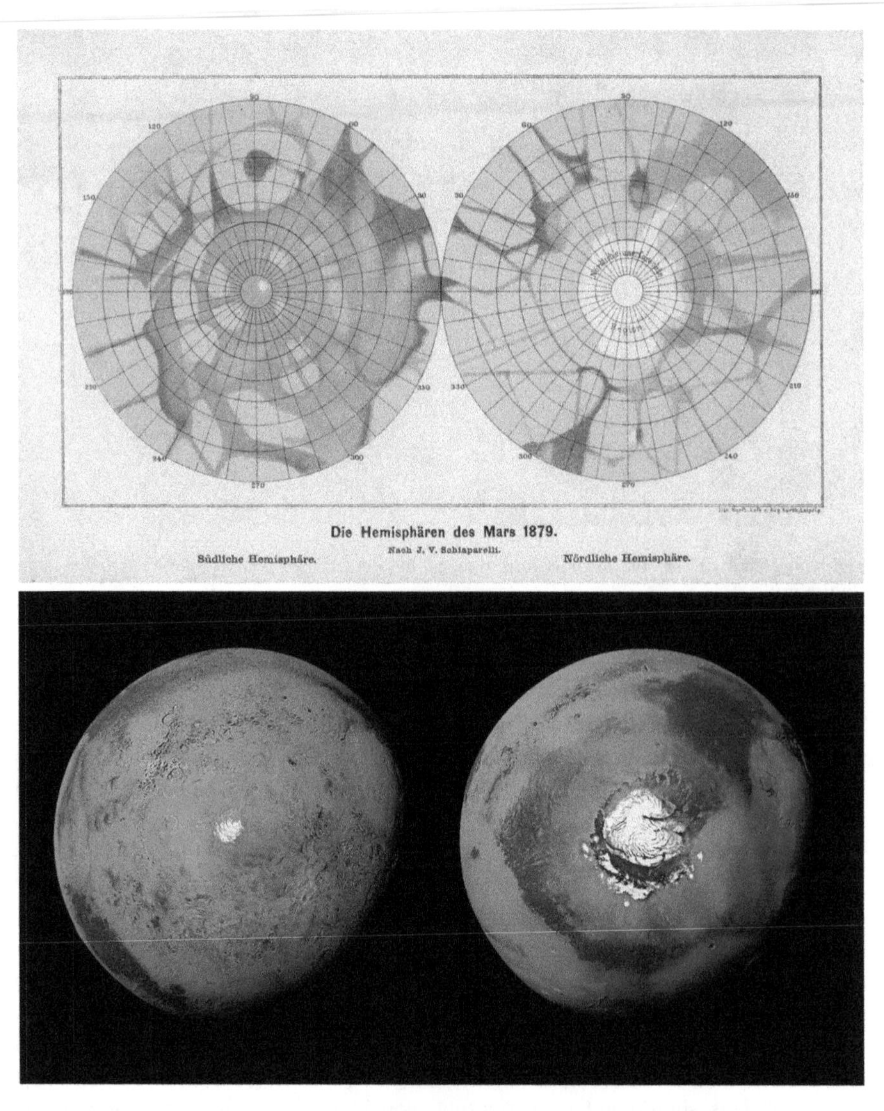

Fig. 4.4 Top: the polar caps were among the first correctly recognized morphologies of Mars. Discovered by Huygens, they were described with precision by Schiaparelli, who recognized the northern cap as the largest. Bottom: a modern image of the Martian poles. Top: image 162368606 FOTOLIA/Archivist, bottom: image 14205030 FOTOLIA/Dottedyeti 0

Glacial and Periglacial Morphologies

On Earth, we know the morphologies that result from ice, even far from the circumpolar areas. There is no need to be a mountain climber to recognize an alpine landscape on Earth. Glaciers, rock glaciers, glacial cirques, moraines, roche moutounée; these are just some of the morphologies that shape mountain areas.

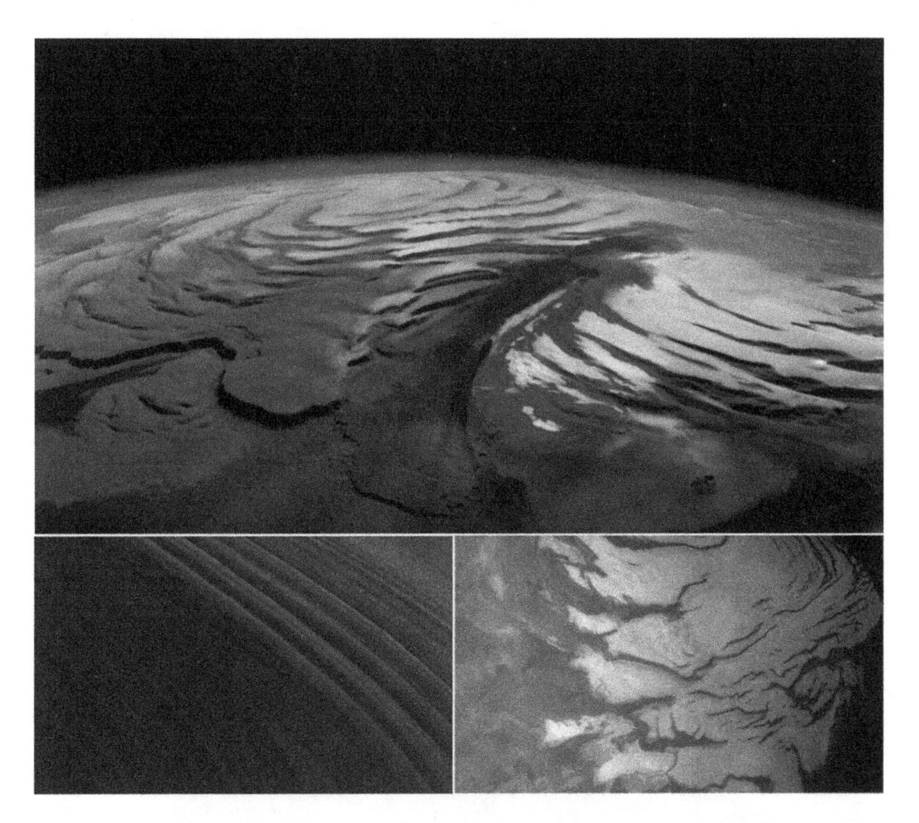

Fig. 4.5 Top: The northern polar cap of Mars. Image ESA obtained with composite HRSC images, superimposed onto MOLA (aboard MGS, NASA) altimeter data. It shows the spiraling arranged ice and the outstanding Chasma Borealis, cutting the polar cap in two. Bottom, left: the north polar layered deposits. The scene, about 3 km across, is part of the HiRISE image ESP_018160_2595. Bottom right: part of the southern polar cap. Image HRSC 231865, courtesy of ESA/DLR/FU Berlin (G. Neukum), CC BY-SA 3.0 IGO. Top: HRSC (Mars Express, ESA) and MOLA (MGS, NASA), bottom left: HiRISE (MRO, NASA), bottom right: HRSC (Mars Express, ESA)

Rock glaciers are a peculiar symbiosis between rock and ice. While the rock provides a dense material necessary for the basal shear stress and simultaneously promotes movement that protects the ice from melting, ice offers the lubricant necessary for a slow but constant flow. The result is beautiful rock tongues. Many of these rock glaciers are fossilized; they date back to the glacial period, and with the advent of the temperate climate, the ice abandoned them, leaving them petrified (Fig. 4.6). Rock glaciers are more significant for Martian study than real glaciers. This is because rock glaciers contain much more solid material, and therefore can potentially show the presence of ice in the past, even after it has disappeared. This is especially useful at lower latitudes than those explored by Phoenix.

That there were morphologies on Mars similar to rock glaciers was known for years before the Phoenix discovery. Figure 4.6 also shows some possible rock glaciers in the southern hemisphere of Mars, just east of Hellas Planitia. The fishbone-like pattern resembles the morphology of a rock glacier. Let us go back

Fig. 4.6 Left: a fossil rock glacier in the Alps. The mass movement, now inactive and colonized by vegetation, took place from the left of the picture. Right: an example of a rock glacier on Mars? The MOC image m0807937 shows part of the rim of a crater at coordinate 39.4 S 247.0 W. Left: FVDB, right: MOC (Malin Space Science Systems, MGS), NASA

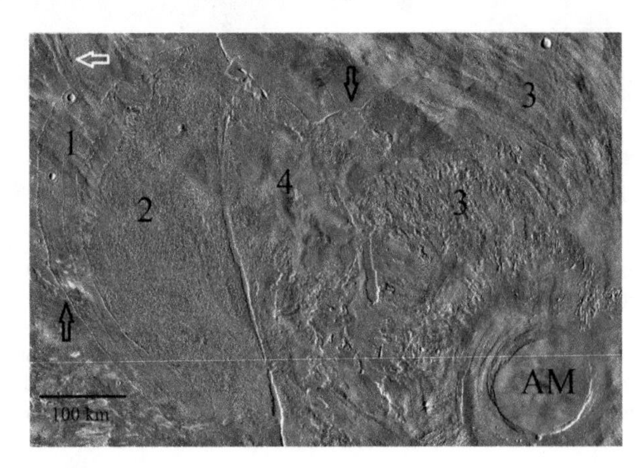

Fig. 4.7 Probable rock glaciers or drop moraines to the west of Ascreous Mons (caldera indicated as "AM") in the volcanic area of Tharsis. The limits of the drop moraines are indicated by black arrows, while a white arrow marks an earlier lava flow that has been successively buried by the moraine. Note the striated areas (marked with the number "1"), which is probably the frontal moraine left by the withdrawing glacier. As opposed to terrestrial moraines, rock material likely derived from volcano eruptions. The other units 2–4 appear different from unit 1, but also belong to the glacier system that departed from the volcano. THEMIS (Mars Odyssey), mdf

to the northern hemisphere. In the northwestern part of the major volcanoes of Tharsis, there are tongues hundreds of kilometers long. Figure 4.7 shows the area west of Ascreous Mons. Here, too, appear probable fossilized ice forms. The striped areas (marked with the number "1" in the figure) are likely moraines left by a retreating glacier. Contrary to terrestrial moraines, rock material was not excavated from the bottom, nor did it fall from the walls of the volcanoes, which are too far

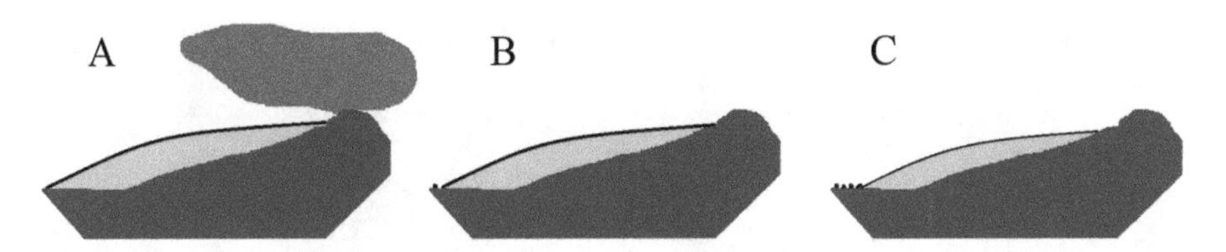

Fig. 4.8 Scheme of formation of a drop moraine. (**a**): an ice lens is buried by airborne material, perhaps pyroclastic ash. (**b**): the ice retreats, leaving a first moraine and then (**c**) more moraines, shown as black dots in the figure. FVDB

away. Rather, it was fine material (ash, lapilli) ejected by the volcano itself. This material slipped to the front by finding its way into deposits called drop moraines. While the 2, 3, 4 units in the figure are a bit chaotic, unit 1 shows beautiful ropes that indicate the ancient glacial positions. As the glacier retreated due to the sublimation of ice, it gave the imprint of its maximum distance. Figure 4.8 shows a scheme of formation of drop moraines: the volcanic ash covering an ice lens (a), which then retreats, leaving successive ropes (b, c). An even clearer example occurs on the northwest border of Olympus Mons (Fig. 4.9). Due to the meteorological circulation system, these structures are typically located on the northwestern sides of the large northern volcanoes. Today, the ice has disappeared. However, a different explanation could be inspired by terrestrial examples of piedmont glaciers (Fig. 4.9, bottom). The Malaspina glacier in Alaska shown here has deposited front moraines during different episodes of expansion and retreat. The difference with respect to the model of drop moraines is that the piedmont glacier model requires flowing ice from the mountain to lower elevations.

Another indication for the presence of subsurface ice is the degradation of ancient craters at high latitude. Figure 4.10 shows a pair of craters within the great impact area of Hellas Planitia. Note the plastic rim shape and the numerous material tongues inside and among the craters, which show antique streams of ice. Crater degradation is even more evident in the example in Fig. 4.11 from Vastitas Borealis in the northern hemisphere of Mars.

SEVENTH MYSTERY: Strange Icy Terrains on Mars

Comparative planetology is the search and description of morphologies that have had similar origin among the terrestrial planets and the moons of the solar system. Impact craters (Sect. 3.1) are a good example, because, although impact velocities on the planetary bodies are variable owing to the different gravity forces on different terrains, the resulting craters are similar on those different planetary bodies. In some cases, however, Martian terrains are so peculiar that there is no obvious counterpart on Earth or any other planet or moon. This makes their interpretation particularly challenging.

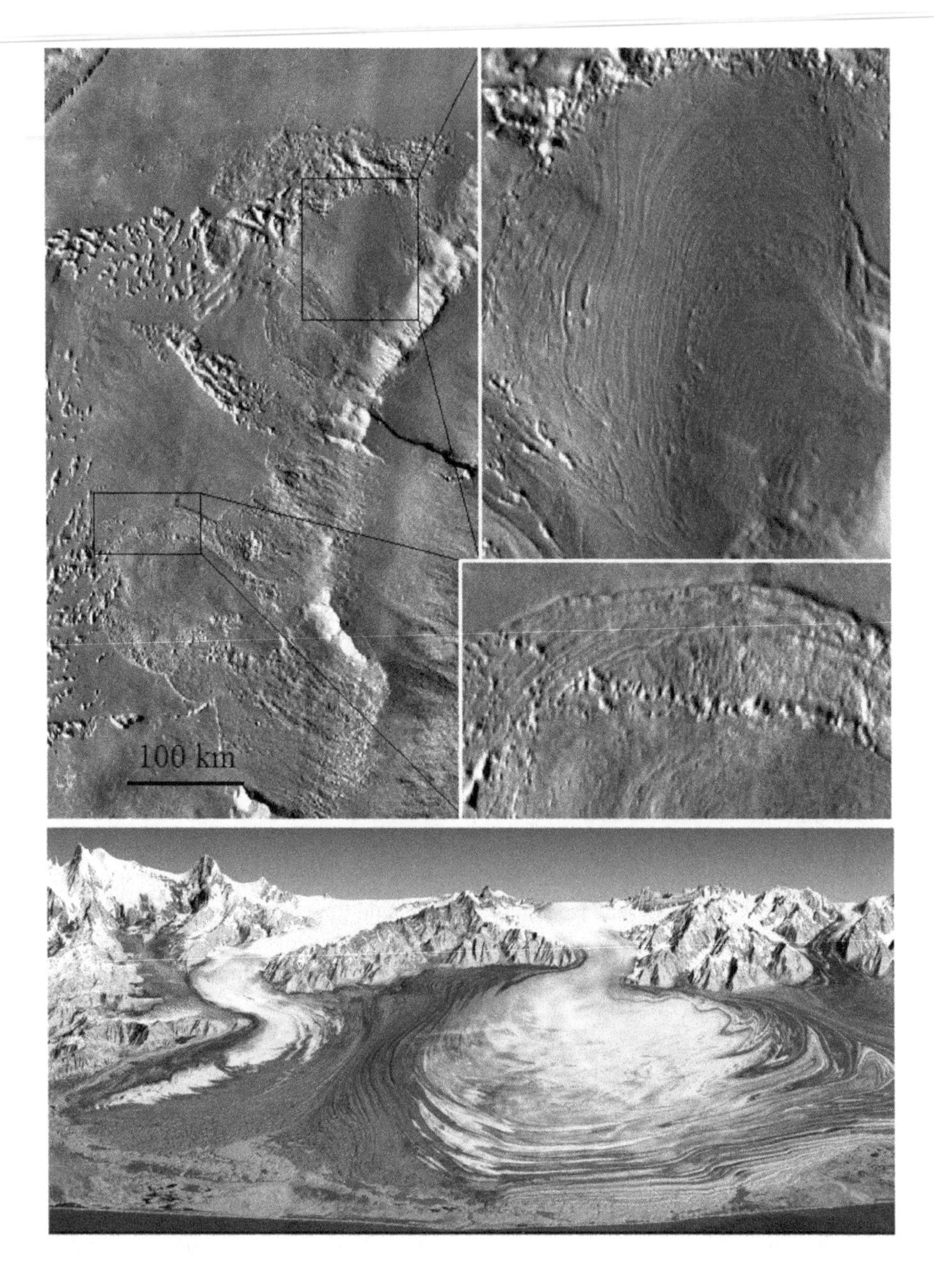

Fig. 4.9 Ancient glaciers in the western part of the Olympus Mons volcano have deposited these concentric lobed traces similar to those of Ascreous Mons in the previous figure. Top left: panoramic view of Olympus Mons. Western shoulder. Top right: two details. These glaciers grew only in the western part of the great Martian volcanoes (all in the northern hemisphere), probably due to the atmospheric circulation system. THEMIS images. Bottom: the Malaspina glacier in southern Alaska. Courtesy LANDSAT/NASA. Top: THEMIS (Mars Odyssey, NASA), bottom: LANDSAT/NASA

Fig. 4.10 A pair of craters at the bottom of the Hellas Planitia impact basin. The scene is about 50 km across. Note the degraded rims, the flow-like structures, one of which appears to have poured off of the larger crater on the left in the figure, and another from the small to the large crater. Image Mars Express 317883, ESA/DLR/FU Berlin. Mars Express, ESA/DLR/FU Berlin

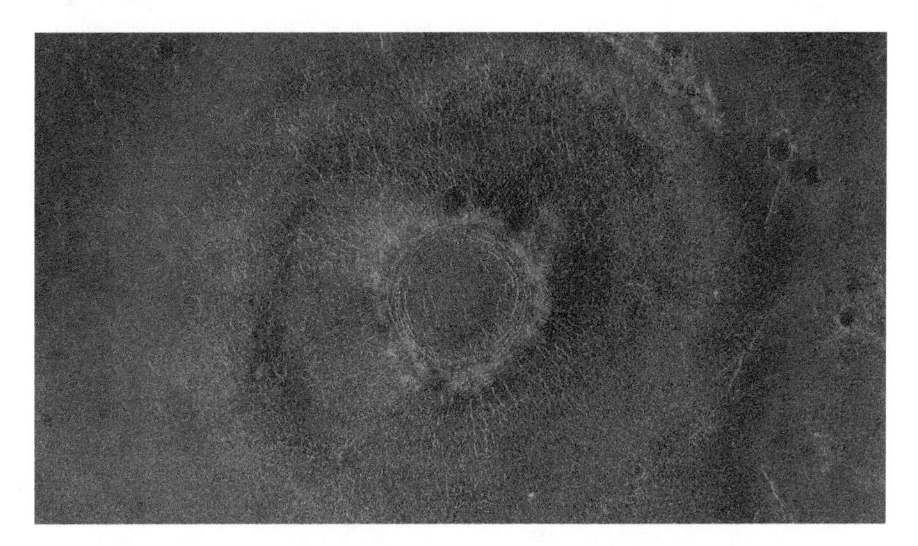

Fig. 4.11 The shape of craters in the Arctic regions has been altered by ice in the ground, like this small crater in the Scandia region. HIRISE_detail_PSP_006825_2465_RED. Scene from left to right about 4 km. HiRISE (MRO, NASA)

Martian chaos terrains are such an example.[1] They consist of a strange network of blocks present in many locations, especially around the region east of Valles Marineris and abounding in the northernmost dichotomy level. Their age ranges mostly from 3.7 to 3.3 billion years, corresponding to the beginning of the Hesperian period to the early Amazonian. Figure 4.12 shows an example of chaotic terrain. In this and many other examples, the terrain seems to have involved an impact crater, visible here below in the image. Within the crater there are blocks separated by a narrow fracture. Craters are often open on one side. In these cases, the material inside the craters seems to have partially escaped through the breach. Outside the craters, the blocks are more separated than inside. Beyond this block zone, there is often a channel, which, in Fig. 4.12, is pointing north. In these channels linked to chaos areas, morphologies resulting from the violent passage of water are evident. In the case of Hydaspis Chaos shown here, there almost seems to be continuity between the crater blocks and those in the channel. Other cases, however, indicate a big difference between blocks as a function of their position. Figure 4.12 shows, on the top right, the exit channel of the great crater of Aram Chaos. Here, one can see the transition between the block crater to the left and the channel to the right, devoid of blocks. In this example, the channel becomes much narrower and shows all of the features of a stream of water from the crater itself. One can also see the erosion conduits inside the crater source. Water seems to have played an important role in the formation of chaotic lands, but today, there is no running water. Other erosion marks are shown in better detail in another chaos area, Iani Chaos (Fig. 4.12, bottom). Where did all of this water come from?

According to an explanation, initially, an impact crater was filled with ice. A certain amount of sediment, always present in the atmosphere of Mars (airborne dust, regolith, pyroclastic material coming from the volcanoes) settled on top of the ice crater lake. But something then heated the ice inside the crater, melting it into water. In some cases, the impact of an asteroid may have been the culprit. Simulations of the heat propagation in the Martian soil after a massive impact show an enduring heat wave. Another source of heat could be magma intrusion or the presence of a volcanic reservoir in the neighborhood. Note that magma intrusion has never been proven on Mars, but there are indications for this phenomenon, widespread on Earth.

However, a simple calculation shows that the heat source could be surprisingly straightforward without the need for exceptional processes. Mars, like our Earth, releases a certain amount of heat across its surface. It is the inner heat of the planet generated by the decay of radioactive nuclides, which, as we saw, gives rise on Earth to convection currents and horizontal plate movement. On Mars, the heat released is presently on the order of 20 mW (milliwatts) per square meter (the heat from a 100-meter-wide ground square would be sufficient to turn on a light bulb). Since radioactive nuclides are decreasing over time, during the Hesperian, the flux was

[1]Because the planetary nomenclature identifies "Chaos" as a terrain of jumbled blocks, there are also features called chaos terrains on Mercury, Europa and Pluto. However, they have a different origin from the Martian examples.

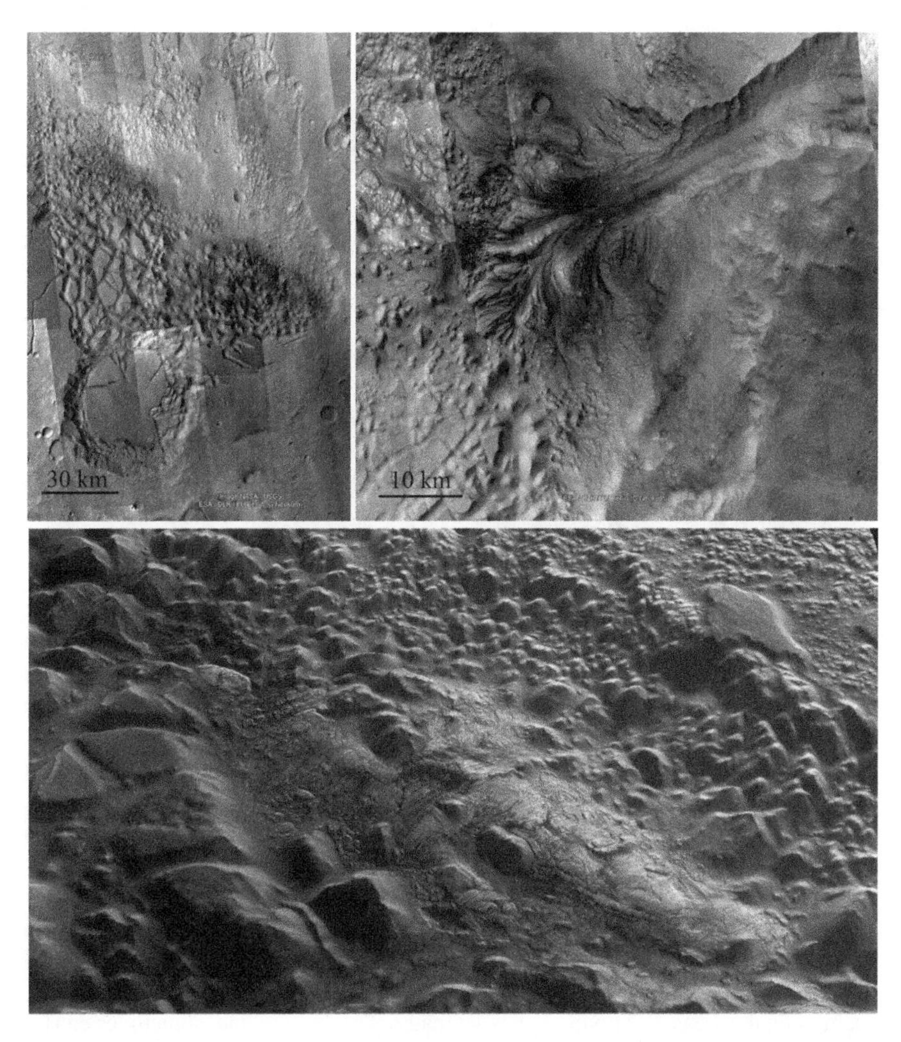

Fig. 4.12 Top left: Hydaspis Chaos. Top right: the eastern part of the huge basin called Aram Chaos. The effect of a water stream capable of carrying huge blocks for kilometers or tens of kilometers is evident. CTX (MRO, NASA) composite image on Google Mars platform; Bottom: more detailed view of Iani Chaos imaged by Mars Express. Image id 220346 about 10 km across. CTX (MRO, NASA)

higher. Even if we cannot be completely sure as to the exact value, a plausible heat flow at the beginning of the Hesperian is between 25 mW and 100 mW per square meter. When there is a difference in temperature between two points of a body, a heat flow is started. If a horizontal bar is heated on one side by a flame, the heat travels from the heated to the cold end. The temperature of the latter will increase until a limit value is reached, in which the heat travelled along the bar is balanced by heat losses. The law of heat flow (also called Fick's law) states that for a homogeneous medium, the difference in temperature between the two ends of the bar is

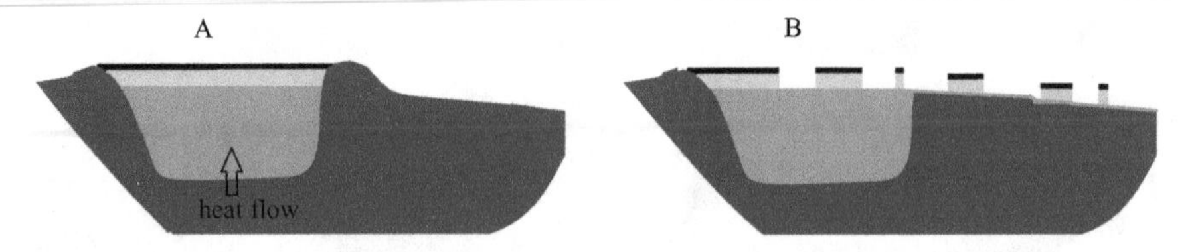

Fig. 4.13 (**a**) Heat flow from below maintains water inside a crater fluid, under a protective layer of ice and debris. (**b**) the rim of the crater is breached, and the consequent forceful water flow rafts the ice-debris blocks outside the crater. FVDB

proportional to the heat flow, multiplied by the distance between the two points of the bar, divided by a material constant of the bar, called the thermal conductivity. The same thing applies to rock. If the heat flow and the thermal conductivity are both known (conductivity for rocks on the Martian surface ranges from 1 to 2, here, we shall use 1.5[2]), we can calculate the temperature rise below the Martian soil. At 1 km of depth, and using the current value for the heat flow of 20, the calculation gives the temperature difference as

Temperature difference $= 0.020 \times 1000/1.5 = 13\ °C$.

Thus, at 1 km of depth, the rocks are about 13° warmer than at the surface. At 3 km of depth, the temperature is nearly 40° higher. The difference was greater in Hesperian times: between 17 and 65° at 1 km of depth for a heat flow of 25 and 100, respectively. This calculation shows that at a certain depth, on the order of 1 km or two, the water may be in liquid form.

There is therefore no need to postulate volcanic or meteorite effects (although in many cases, these may have played a role in melting permafrost) for the obtainment of a 3-km deep crater lake with liquid water at the bottom. According to this model, a deep crater lake had three layers: a basal water layer and an intermediate ice layer, all coated with a layer of rock and regolith, here referred to loosely as "debris" (Fig. 4.13). We know that on terrestrial glaciers, water pools are often formed within or at the sides of the glacier. These lakes are sometimes confined by the lateral moraine, that is, the solid material transported to the sides by the glacier itself. Water, pushing on the weak and poorly cohesive material of the moraine, sometimes causes its collapse. In mountain areas, this sudden rupture of confining ice sometimes forms a dangerous, unusual stream of debris, ice, and water called GLOF (Glacial Lake Outburst Flood).

Something similar, although on a much greater scale, occurred in these chaotic terrains. The water inside the crater exerted a heavy pressure on the rim, causing its collapse. Leaving the side of the collapse, the mixture of water and ice opened the way to the broken blocks, which were transported as rafts in the surrounding area. Blocks were then abandoned by the water flow, which continued its path and eroded the soil, creating the flow channels. If the channel was too narrow, like in Aram

[2]As usual, we shall be using international units, and when appropriate, we will not report the units in the main text for simplicity. Conductivity has units W/(°C M), heat flow of W/m^2.

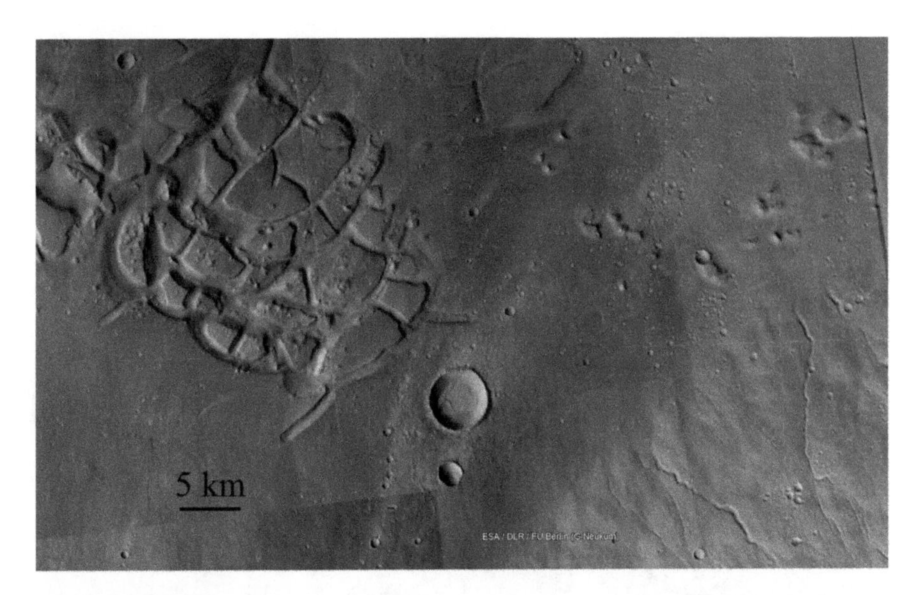

Fig. 4.14 Detail of blocky terrain in Galaxias Chaos. Bottom right is Elysium Mons. CTX mosaic on Google Mars platform. CTX (MRO, NASA) mosaic on Google Mars platform

Chaos, the blocks were not transported much, but remained inside the crater. However, in the case of Fig. 4.12 (top left), it is clear that the area of the blocks involved is much larger than the size of the crater.

A different hypothesis suggests that rafts were carried not by pure water, but by a much denser debris flow. The advantage of this hypothesis is that a debris flow is a water–rock mixture, and thus has the capability of transporting heavy tabular slabs better than water. Furthermore, it does not require much water to flow. One problem with this explanation is the presence of channels beyond the chaotic terrain. Although it has been suggested that they may have been carved by debris too, water appears to be a more plausible explanation, since debris flows tend to accumulate, rather than erode, in flatter areas.

It has been noticed that not all chaos terrains are alike. In Galaxias Chaos, north of Elysium Mons, blocks are not associated with craters, and the outflow channels are narrower and more distant (Fig. 4.14). Thus, two of the three characteristic morphologies of the "classical" sequence of chaos terrain are absent. The valid explanation for these terrains must be different, and probably involves the lava flows from Elysium Mons.[3] Thus, it appears that the Martian Chaos terrains may have more than one single explanation, but all require a mobile medium underneath the blocks.

Another type of enigmatic terrain is shown in Fig. 4.15. It shows strange blocks a few miles across against a lighter background. We are in the equatorial region of Elysium Planitia, a few degrees north in latitude (7°). Like the pieces of a giant

[3]Pedersen, G. B. M., & Head, J. W. (2011). Chaos formation by sublimation of volatile-rich substrate: Evidence from Galaxias Chaos, Mars. *Icarus*, *211*(1), 316–329.

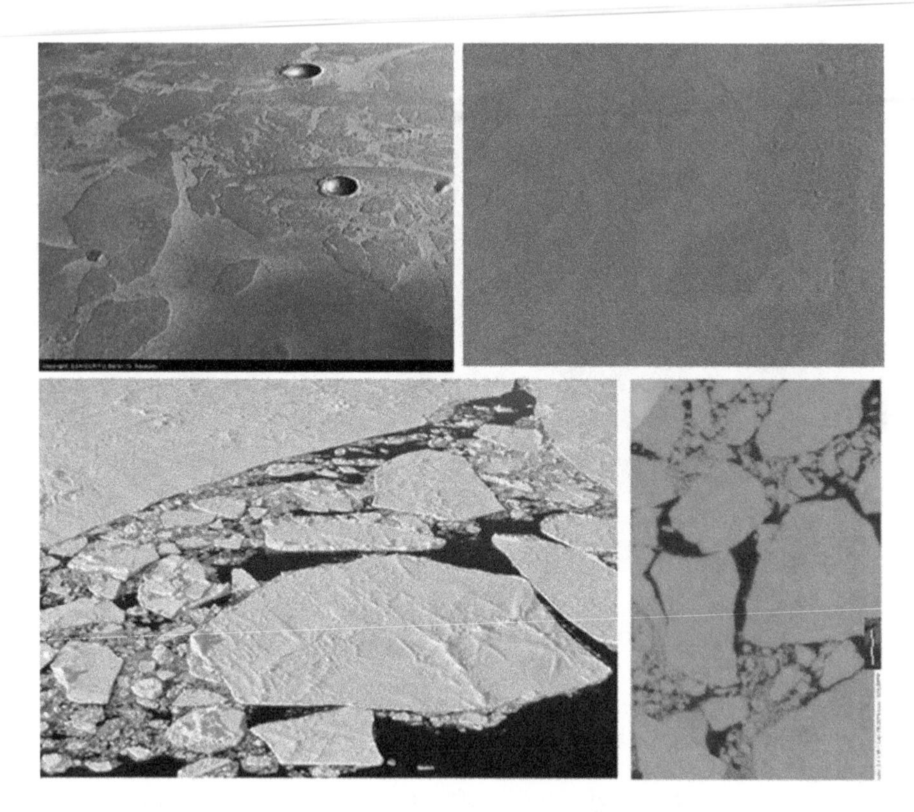

Fig. 4.15 Top left: a well-known HRSC image of blocks in Elysium Planitia from Mars Express. The scene is about 30 km wide. To the right: more blocks imaged by the CTX camera. Bottom, left: blocks in the Arctic ocean. Bottom, right: LANDSAT image from Prince Patrick Island in the Beaufort Sea, taken on August 19, 2010. Bar on the right 6 km long. Courtesy of USGS. Top left: HRSC (Mars Express, ESA), top right: CTX (MRO, NASA), bottom left: image 105501725 FOTOLIA/Vladimir Melnik, bottom right: LANDSAT, USGS

jigsaw puzzle, these blocks once formed an unbroken structure. What did this structure consist of, and why did it break apart, leaving blocks to drift away from each other? One initial explanation is that these are solidified blocks of lava. We know that some basaltic volcanoes, such as Kilauea in Hawaii, form lava lakes that harden at their surface. However, solidified lava is heavier than fluid magma and tends to fall into the magma lake. A second explanation is that blocks are chunks of thin ice. Being lighter than water, ice does not sink. The next two figures (Fig. 4.15, bottom) provide an eloquent comparison with the icy terrestrial sea. The first one shows small ice blocks in an Arctic bay, the second one a satellite image of large blocks in the polar Beaufort sea. The blocks' size is comparable to that of those on Mars. Also, in these images, it is easy to reconstruct the relative movement of each block. Notice that the Martian blocks in Fig. 4.15 (top, right) seem more irregular and richer in craters compared to the surrounding clear zone. This is well explained if the blocks were ice floating on an ocean. The bottom of the ocean was spared somewhat from the meteorite impacts, partly due to the shielding effect of water and

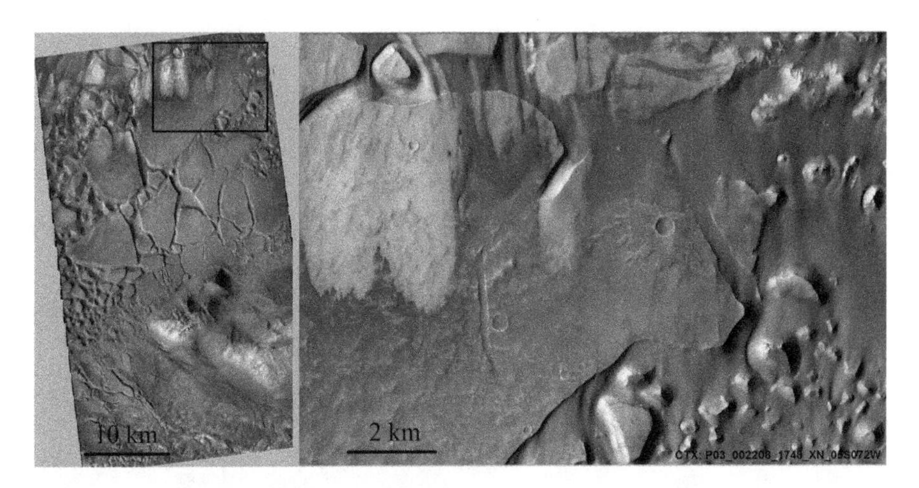

Fig. 4.16 Blocks in Candor Chasma. CTX. Left: general view. Right: magnification of the area in the rectangle with a crater with rampart ejecta. CTX (MRO, NASA)

partly because of resurfacing at the sea bottom. Probably, the Martian blocks were covered with clay or sand transported by the wind, otherwise, there is no explanation as to how the traces persevered over time. Thus, this might be yet another indication in favor of an ancient Martian ocean.

Figure 4.16 shows other similar blocks in Candor Chasma, within Valles Marineris. The blocks almost seem to follow an ancient flow into a water basin directed toward the bottom of the figure. A more detailed view of one of the blocks shows the presence of a rampart crater, typical of impacts on icy soil. Does this confirm that these were actually ice blocks?

4.2 Other Indications of Icy Mars

Water and Volcanoes

Figure 4.17 shows morphologies reminiscent of giant frozen bubbles geometrically oriented along apparent flow lines. Seen closely, each bubble looks like a crater bulging from the ground. We know how these craters form, because there are identical landforms in Iceland. Often called rootless cones, they are created when hot lava flows on top of wet or frozen soil. In contact with the high temperature of lava, the water underneath vaporizes promptly. Because vapor, like any other gas, expands dramatically with increasing temperature and, at the same time, increases in pressure, it punctures the overlying lava layer, creating a shallow cone. The orientation of rootless cones indicates the direction of the lava flow. Thus, their presence on many Martian terrains is proof that some lava was most likely flowing over an icy or water-rich area. Rootless cones on Mars may develop fields with a pattern controlled by the orientation of the lava flow, creating picturesque forms. Figure 4.18

Fig. 4.17 Above: Particular, patterned lava flows such as these are found in many areas of Mars, especially in the northern hemisphere. This CTX image shows an area around Phlegra Montes. Lava traveled from the top right to the lower left. The "volcanoes" well visible here, properly called "rootless cones," are well known from Iceland (below). They form when lava flows over a damp or icy area, thus vaporizing the water. High pressure steam, rising up a few tens of meters of lava, reaches the surface in the form of small vapor eruptions. Top: CTX (MRO, NASA), bottom: image 144042166 FOTOLIA/verve

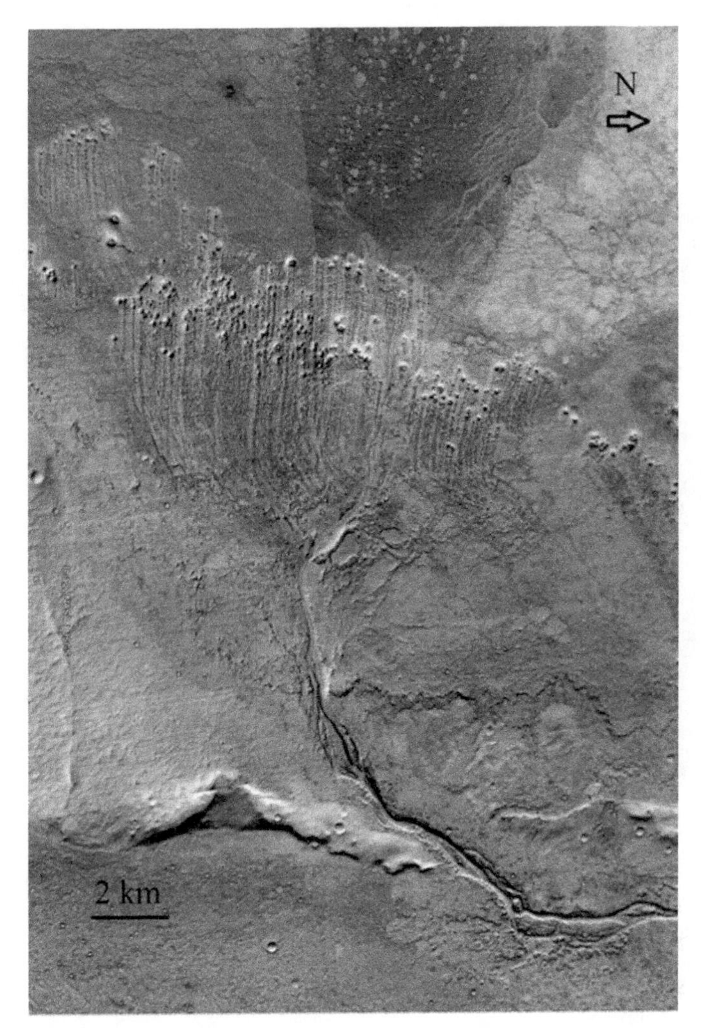

Fig. 4.18 Rootless cones 25° N 54′, 173° 52′ E, east of Phlegra Colles. To accentuate an amusing resemblance with a bush (or perhaps even better with a fossil crinoid), north is to the right. CTX (MRO, NASA), mosaic on Google Mars platform

shows the result of a lava flow coming from the right (corresponding to the north in the figure) and then changing direction to form a "stem," before flowing onto an icy surface and drawing branches of a sort.

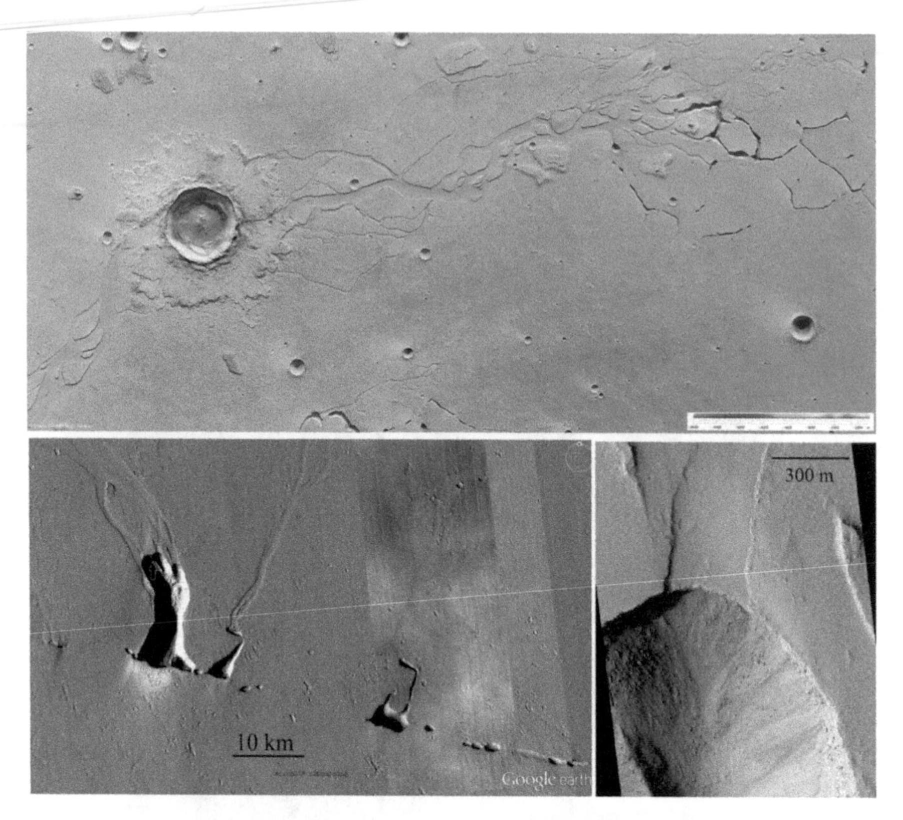

Fig. 4.19 Top: One crater in Hephaestus Fossae (20° 10′ N; 126° 03′ E) exhibits layered ejecta and a waterway network. HRSC elevation colorized map. North points to the right. Scene from left to right about 160 km. Bottom left: similar waterway departing from the Tooting crater in Amazonis Planitia. CTX mosaic on Google Earth platform. Bottom right: detail of the previous image in correspondence with the white arrow. It is likely that a system of caverns similar to the ones on Earth have been created along the fracture. HiRISE image. Top: HRSC elevation model (Mars Express, ESA), bottom left: CTX on Google Mars platform; bottom right: HiRISE (MRO, NASA)

Layered Impact Craters and Landslides Again: Yet Another Clue to an Icy Mars?

We now reconsider two landforms seen earlier: layered ejecta around impact craters and landslides. We saw that layered ejecta are observed only around craters of sufficiently large diameter. If ice is responsible for the creation of the layered ejecta, their absence around small craters can be understood by considering that the small craters are also shallow, and therefore cannot reach the deep ice layer. A large enough meteorite, however, may create a hollow during the excavation phase deep enough to involve subsurface ice. The ice inside the pores melts into water and then steam, fluidizing the ejecta in the petal-shape so characteristic of the layered ejecta.

As evidence for this process, we can refer to Fig. 4.19. It shows a layered crater in the Hephaestus Fossae area from which two alleged waterbeds depart at opposite sides. Tens of kilometers away from the crater, the waterbeds branch. The story of this amazing morphology is as follows: the fall of the meteorite that has produced the crater has excavated into the ice layer, heated the terrain to a certain depth, and melted conspicuous amounts of ice. We know that temperatures produced at the surface during a large meteorite impact are far beyond the melting point of water, and thus a heat wave propagates down into the planetary surface for several kilometers. The warm conditions in the deep soil may also persist for several thousand years or even a million years for very large impactors. As a consequence, hot meltwater pouring out of the impact area has incised the channels in the figure, creating the complicated water erosion pattern. The channels' erosion was thus not merely a mechanical process, but probably also involved a softening of the soil and ice melting of the ice-bearing soil along the path. Note that in finding a deep fracture, water has also found a preferential path that penetrates into the ground. A similar and even more evident process has occurred with the Tooting crater in Amazonis Planitia (Fig. 4.19). Here, two separate flows have come across a deep and very long fracture (which continues underneath the Olympus Mons aureole to the east) and directly penetrated into the deep ground, where the water, perhaps still hot, melted the subsurface ice. The local scene, taking place over a short period of time, must have been breathtaking: large meteorite fall, temporary rivers developing and bringing waters several tens of kilometers away, water seeping along the ten-kilometer-long fractures in the terrain, waterfalls at the intersection with fractures, where deep caves were excavated! The waterfalls retreated with time, a process that gave origin to the deep ditches at the boundary between the watercourse and the fracture (one of which is enlarged in the figure at the bottom right).

Landslides have a dynamic behavior somewhat similar to ejecta. They also mobilize a deep layer of Martian surface, and on Mars, they probably travel at high speeds (the largest at speeds even greater than 100 meters per second, or greater than 360 kilometers per hour!). If we now compare some of the landslides in Valles Marineris (Fig. 4.20) with those that fall on terrestrial glaciers (Fig. 4.21), we notice some remarkable similarities. Both are reminiscent of a viscous liquid that expands both sideways and along the direction of the main slope. The closest analogy, however, is in the longitudinal grooves that accompany both terrestrial and Martian landslides along their path. What is the reason for the resemblance of Martian landslides with those confined to glaciers? We don't see any superficial ice today underneath the landslides, because Valles Marineris is at the equator. Thus, if ice is the correct explanation for this resemblance, either there was superficial ice early in the history of Valles Marineris or the landslides carried with them a large amount of ice initially present in the pores. Others consider that landslides travelled lubricated by clay, rather than ice. Some landslides have been dated by various researchers. While most landslides are recent (less than 500 Ma), there are also examples of much older ones. The oldest dated on Mars have an age of 3.5 Ga, i.e., the early Amazonian. Understanding the dynamics of these landslides can shed light on the state of Mars' surface at the time of their collapse.

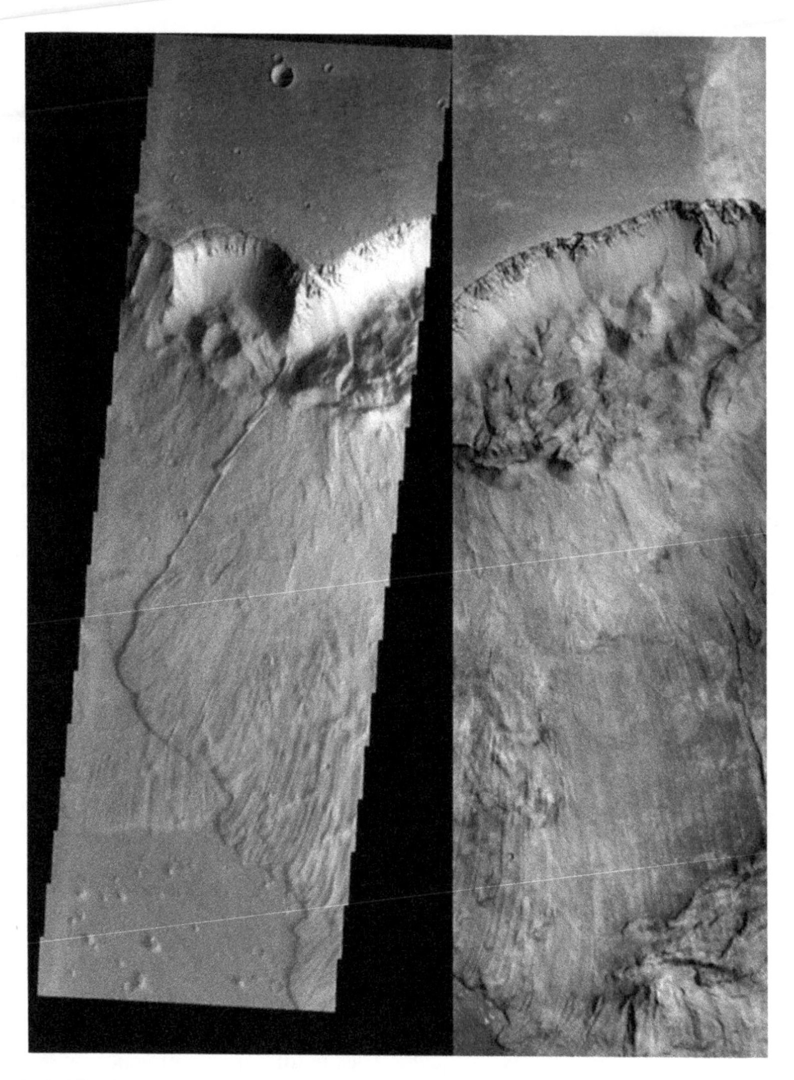

Fig. 4.20 The landslides shown here are in Ganges Chasma of Valles Marineris. Left: Two landslides depart from their respective scars in the form of amphitheaters. The landslide on the right has collapsed later and partly covered the previous one. Image Themis PIA09057. B: Image of the same landslide. Image Themis 20020401, coordinates −8.6 N, 315.7 E. The purpose of these figures is to show the streaks, to be compared with those shown by glacial landslides in the following figures. THEMIS (Mars Odyssey, NASA)

Technical Box 7: Water Map from Gamma Ray Spectrometry

The optical images of spacecraft in orbit around Mars are particularly familiar, because they are centered in the electromagnetic field to which our eyes are

(continued)

Technical Box 7 (continued)
sensitive. In fact, there is a lot of other information that has been acquired by remote sensing using different means other than optical photography.

A refined technique has been used in the exploration of the Martian surface. As far as we know, there can be no life without water. Thus, revealing the presence of water is a fundamental task in Mars exploration. Gamma ray spectrometry (GRS) is based on the principle that a neutron (neutral particle, one of the two kinds of particle composing the atomic nucleus, the other being the positively charged proton) hitting a hydrogen nucleus (composed of a single proton) emits a characteristic photon, a gamma ray of energy 2.223 MeV (MeV is the unit of energy used in nuclear physics; it corresponds to $1.6 \cdot 10^{-13}$ J). Because water is composed of two atoms of hydrogen, the presence of water can be detected by such gamma rays at the expected energy. Luckily, we do not need to irradiate the Martian soil with neutrons, which would be impossible from a far-away spacecraft. Cosmic rays from the sun and from outside the solar system continually impinge on the Martian surface, cracking the nuclei of various elements, and thus liberating a cloud of neutrons that eventually impact on the hydrogen. With this technique, the map in Fig. 4.22 has been obtained by the Gamma Ray Spectrometer at Mars Odyssey. The device is composed of a scintillator that transforms gamma rays into light, and then converts light into an electrical output with a photomultiplier tube, a standard technique in nuclear physics. When neutrons impinge on atomic nuclei other than hydrogen, a different gamma-ray signature is produced, which can be identified. Thus, gamma ray spectrometry can provide the elemental composition on the top decimeters of the Martian surface. The following map in Fig. 4.22 shows the superficial distribution of water. Notice the transition at about 45° latitude from a dry to an icy terrain. For analysis of the deep terrain, other techniques such as radar have been used.

4.3 Watery Mars

Outflow Channels

Numerous channels join the southern highlands to the northern plains. Some of these channels, which have no correspondence to the ones drawn by Schiaparelli and Lowell, are shown in Fig. 4.23. They are several hundreds to thousands of kilometers long. While some seem to start from a relatively open basin (e.g., Kasei, Shalbatana), others have an apparently sudden origin. One outstanding case is Ravi Vallis (Fig. 4.24). It apparently starts from amphitheater-like morphology more akin to a landslide than a river, developing deep basins alternating with visibly eroded channels, and finally plunging onto chaotic terrains.

Fig. 4.21 The comparison of landslides on terrestrial glaciers with the Martian landslides shows a remarkable resemblance. Left: Sherman's great landslide in Alaska due to an earthquake in 1964. It is flat, and the streaks make it similar to many Martian landslides. To the right: another landslide on an Alaskan glacier: the recent de la Perouse landslide. The noticeable curvature at the front of the landslide indicates that, at this point, it was moving at low speed. Photo courtesy of FlyDrake Haines Alaska. Left: USGS, right: courtesy FlyDrake, Alaska

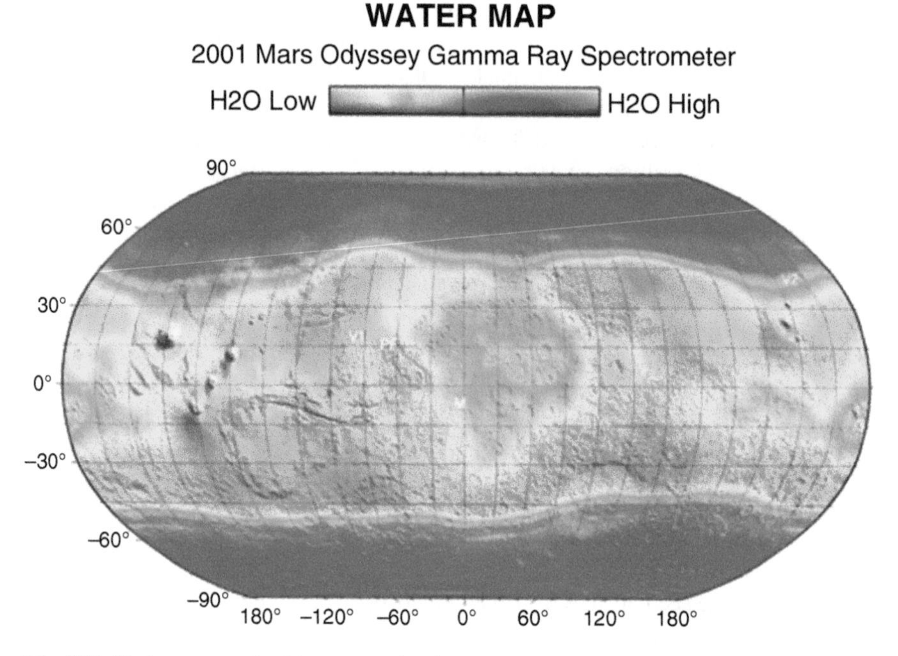

Fig. 4.22 Hydrogen contained in water molecules on the surface of Mars reconstructed by gamma ray spectroscopy. Note that superficial water (which is in the form of ice) is abundant for latitudes higher than about 45°. Image Credit: NASA/JPL/UA. Gamma Ray Spectrometer (Mars Odyssey, NASA)

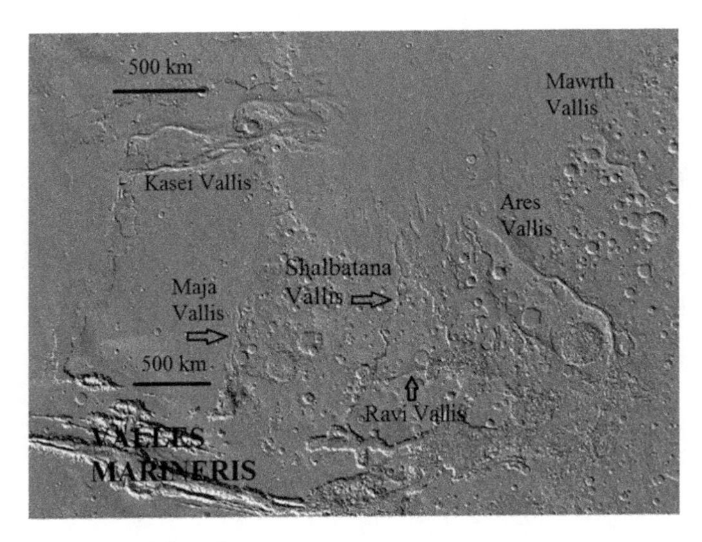

Fig. 4.23 Outflow channels (Valles) northeast of Valles Marineris. The different length scale is due to cylindrical projection. MOLA (MGS, NASA)

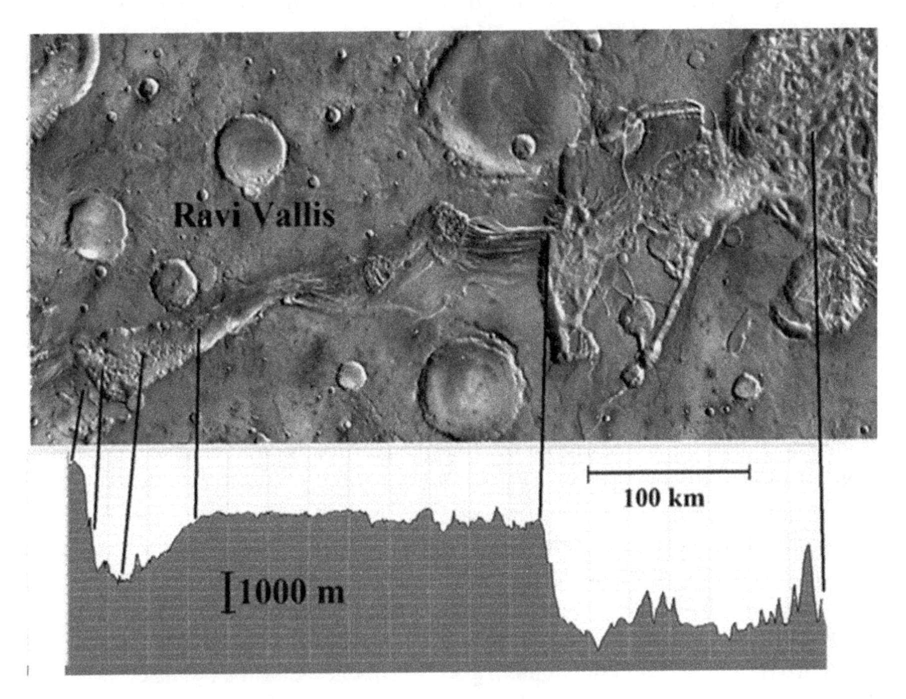

Fig. 4.24 Above: THEMIS map of Ravi Vallis (head of valley 0.8° S, 43.1 W). Below: the altimeter profile MOLA reveals that the head of the valley occupies a depression. The vertical lines make a correspondence between the planimetric map and the altimetry. The basin at Ravi Vallis's head must have been filled for nearly 2 km. THEMIS (Mars Odyssey, NASA) and MOLA profile (MGS, NASA)

Fig. 4.25 Kasei Vallis in a THEMIS image. THEMIS (Mars Odyssey, NASA)

Let us now look closely at another channel: Kasei (Fig. 4.25). The set of channels, the sinuous shapes, the craters surrounded by drop-shaped islands, all suggest erosion by an energetic watercourse. This image was taken near the border between the southern highlands and the northern plains. The water flowed from south to north, perhaps filling or shrinking the great ocean of Mars. Was it a stable river or a temporary catastrophic watercourse?

The deep erosion typical of these channels and the low number of tributaries (terrestrial rivers have many more tributaries) exclude these from being "normal" rivers fed by precipitation, but at the same time, can drive our intuition as to the origin of such channels. Earthly rivers, which are usually quite stable, may sometimes increase in discharge and speed, resulting in floods. One particular kind of flood that occurs in glacial environments has been introduced earlier in relation to the chaos terrains.

Floods are certainly adverse phenomena affecting human settlements, but are nothing in comparison to some catastrophic drainage that has occurred in the past. With glacial melting at the end of the last glacial expansion, a lake was created in North America in the present state of Washington (USA), called Missoula by geologists. The huge water mass was confined within natural ice walls. Approximately 8000 years ago, it emptied forty times, releasing volumes of water so enormous that very wide channels were carved in a geologically short time, and erosion tens of meters deep created the "Channeled Scabland" (Fig. 4.26).

Other forms of water erosion are shown in Fig. 4.27. The gorge that cuts the image from right to left in the figure is part of Coprates Chasma (Valles Marineris). Among the multiple fractures, quite typical around Valles Marineris, the ones closest to the valley axis have been partly eroded and widened. Such morphologies are reminiscent of sapping, a form of erosion initiated by groundwater. Examples of

Fig. 4.26 Top left: the Palouse River and (right) the Palouse Falls are part of the Channeled Scablands (two figures below), created by the emptying of the Missoula lake. The Channeled Scablands are considered a good analog to the Martian outflow channels. Top left: image 125042069 FOTOLIA/Zack Frank, top right: image 125042160 FOTOLIA/Zack Frank, bottom left: image 125042528 FOTOLIA/Zack Frank, bottom right: image SHUTTERSTOCK

sapping in the Grand Canyon are shown in Fig. 4.28. On Earth, sapping occurs when water springs out at the level of the water table and erodes the terrain. Then, the erosion proceeds away from the main valley. The results of such erosion are remarkably similar to Martian examples, like the ones shown in the figure.

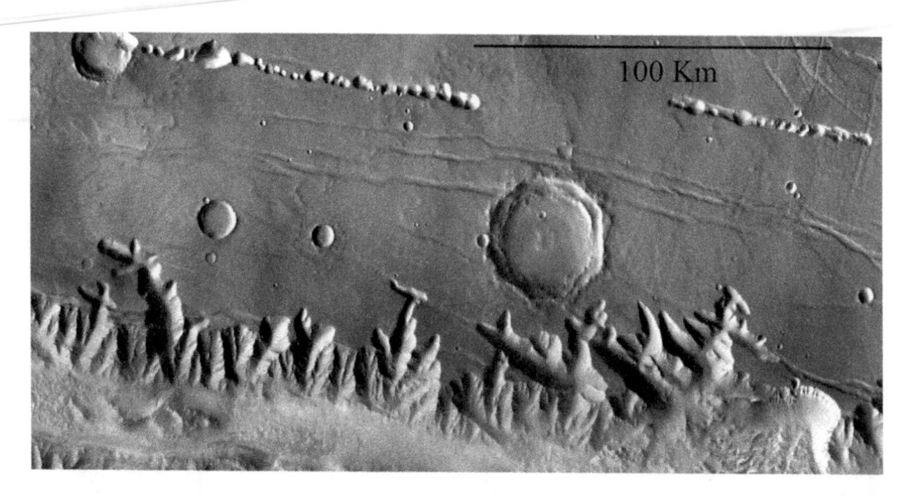

Fig. 4.27 Possible erosion forms due to water (groundwater sapping) are visible at the bottom, started in correspondence with fractures in Ius Chasma. Note how some of the gullies where sapping has started belong to swarms of pre-existing fractures. Note also the pit chains at the top. Image THEMIS daytime. THEMIS (Mars Odyssey, NASA)

Fig. 4.28 Sapping erosion in the Grand Canyon. Image 61941339 FOTOLIA/fannyes

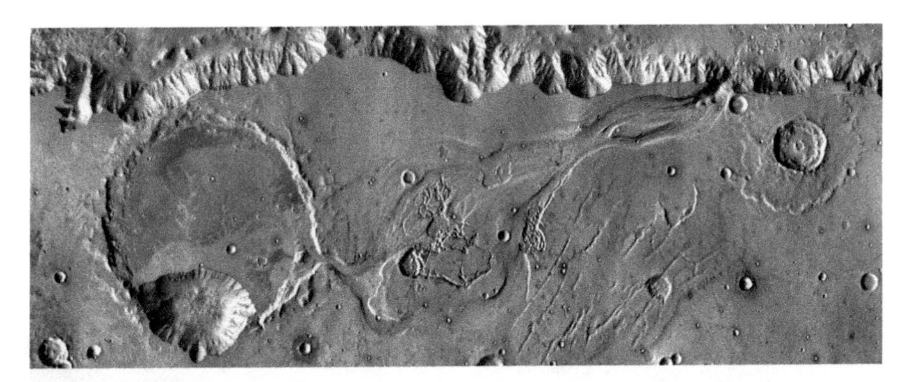

Fig. 4.29 A complex story exists for the water flow from this crater south of Coprates Chasma. THEMIS (Mars Odyssey, NASA)

Lakes?

There is yet another way to detect the presence of running or standing water in the past, which is used by geologists to assess ancient lakes on our planet. It consists in the study of sediments transported and deposited by water, rather than erosion patterns. One amazing story of water erosion–deposition is narrated by the infrared image in Fig. 4.29, showing a large crater on the left and a multiple drainage channel extending eastwards, a typical erosion channel. Originally, a lake filled the impact crater visible in the figure. The presence of water is also demonstrated by the fractures of the dissection, the "mud cracks" inside the crater (we shall find out more about mud cracks as a proof of drying water). Water had to come from the crater itself, as there are no external sources. One day, an unknown event breached the crater rim, causing eastward drainage and channel excavation. But it did not occur all of a sudden: examining the channel closely, we can see how water flow has migrated over time to find the best gradient path. The flow went straight into Coprates Chasma to the north (top of the figure), also forming a depositional delta at the slope break. Then, in the southern-facing part of the crater, there occurred a collapse that formed a 3000-meter-deep pit.

Other stable lakes have been suggested for impact craters on Mars, such as the massive impact basin of Hellas Planitia, in addition to several craters. Valles Marineris itself shows the presence of sections of sediments interpreted as deposited into an ancient lake. The "ILD," standing for "Interior Layered Deposits," is a peculiar series of layers observed in some locations of Valles Marineris, especially in closed Chasmata such as Hebes Chasma. It is regarded by some as the product of lacustrine sedimentation in Valles Marineris. The lowermost location in Valles Marineris is reached in the area of Melas-Coprates Chasmata. According to some researchers, a lake was present in these depressions some time during the late Hesperian-early Amazonian. The lake then breached, and huge amounts of water were released to the north.

Fig. 4.30 A Martian rock on the left, as seen by the Curiosity rover. The resemblance to a terrestrial clastic rock with river-transported rounded gravel (two images on the right, taken in Sardinia) is evident. All three images taken in Summer 2012. Left image pia16189-43_full, courtesy of NASA. Left: Curiosity image (JPL, CALTECH, NASA), top and bottom right: FVDB

On Earth, sediments transported by rivers are characteristically made up of clasts and blocks of rounded shape acquired after tens of kilometers of transport. Unfortunately, no scientist has ever examined a section of Martian river sediments, but the Curiosity rover, which landed on Mars in August 2012, did. Figure 4.30 shows a photo of river sediments as the rover saw them. They look like those on Earth.

Desiccation Cracks

When a lake dries out, the clay on the surface loses water and shrinks. Contracting clay forms artistic tabular blocks separated by cracks (Fig. 4.31). Geologists have often used these structures to infer an arid climate in the study of ancient sediments. The figure on the right shows ancient mud cracks from the Permian period, demonstrating that this part of the Alps was invaded by intertidal waters that could form temporary ponds.

Identical structures have been observed on Mars. Figure 4.31 shows an amazing picture taken from the Curiosity rover. This rover, one of unprecedented technical

Fig. 4.31 Left: mud cracks are shrinkage structures in a clayey soil; they form when a pond or lake dries out. Death Valley, California. Right: when similar dehydration structures are found in sedimentary rocks, geologists deduce a paleo-environment of stagnant or lagoon waters subject to

capabilities, has been examining the Gale crater and its neighborhood in great detail, using both imagery and refined chemical and nuclear analysis. Gale is a 150 km-wide crater in Aeolis Mensa, south of Elysium Planitia, characterized by a high central mound (called Aeolis Mons) that is, in fact, higher than the terrain surrounding the crater. These polygons were found at Mount Sharp, along the planned path of Curiosity. Desiccation cracks confirm that 3 billion years ago, there was a lake here, which dried out.

EIGHTH MYSTERY: Water Riddles on Mars, or the Nature of Outflow Channels and the Vastitas Borealis Formation

Figure 4.32 shows another channel system and erosion in the terminal part of Ares Vallis. The altitude profile (figure below) shows that the depths reached by the gorges varies between three and four hundred meters, while the width of each sub-channel is about 20–50 km. However, the depth of the gorge records the total erosion by the Martian waterways, and is not necessarily related to the height of the water in the channel. Was the channel dug during a brief episode of enormous flow or during the slow but steady stream of water?

The fact is, we do not know the answer, but we can make estimates based on well-established laws of water motion in a channel. While some practical notions of hydraulics were well-known to the Romans, who needed water in the big cities and built aqueducts with precise slopes to make water flow at the right speed, it was only in the eighteenth century that French and then German engineers began to develop quantitatively the field of hydraulics. Today, we can rely on the so-called Manning's law, which allows us to calculate the speed based on the height of the free water level and the slope of the water course. To take into account the water's friction with the bottom of the channel, a coefficient of Manning is introduced, which depends on the ruggedness at the base of the watercourse. The speed increases with water level and gravity acceleration. For example, the velocity of water in a channel is six and a half times greater when the water level is 100 m rather than 10 m. A great difference is reflected in our ignorance of the discharge of these streams (discharge is the volume of water brought in a second through a section of the watercourse). Estimates vary, depending on whether the author has catastrophic or more gradualistic visions of these phenomena. For Kasei Vallis, proposed discharges vary from ten thousand cubic meters per second (even less than the flow of the Mississippi) to one cubic meter per second, which is a hundred times less than the estimated peak for the Scablands. Based on Manning's law, at maximum capacity, the "Kasei River" had to run at speeds between 6 and 7 meters per second, or between about 21 and 25 km/h.

Fig. 4.31 (continued) episodes of dryness. The cracks shown here date back to the Permian and are 250 million years old; vertical height about 1 m. Middle: the "Old soaker" slab, as seen by the Curiosity rover (the name comes from an island in Maine). The whole picture is about 2 m across. Lower image: a detail of the previous image. Top left: image 169865850 FOTOLIA/Ronald, top right: courtesy R. Gualdi editore, middle and bottom: Curiosity image (JPL, CALTECH, NASA)

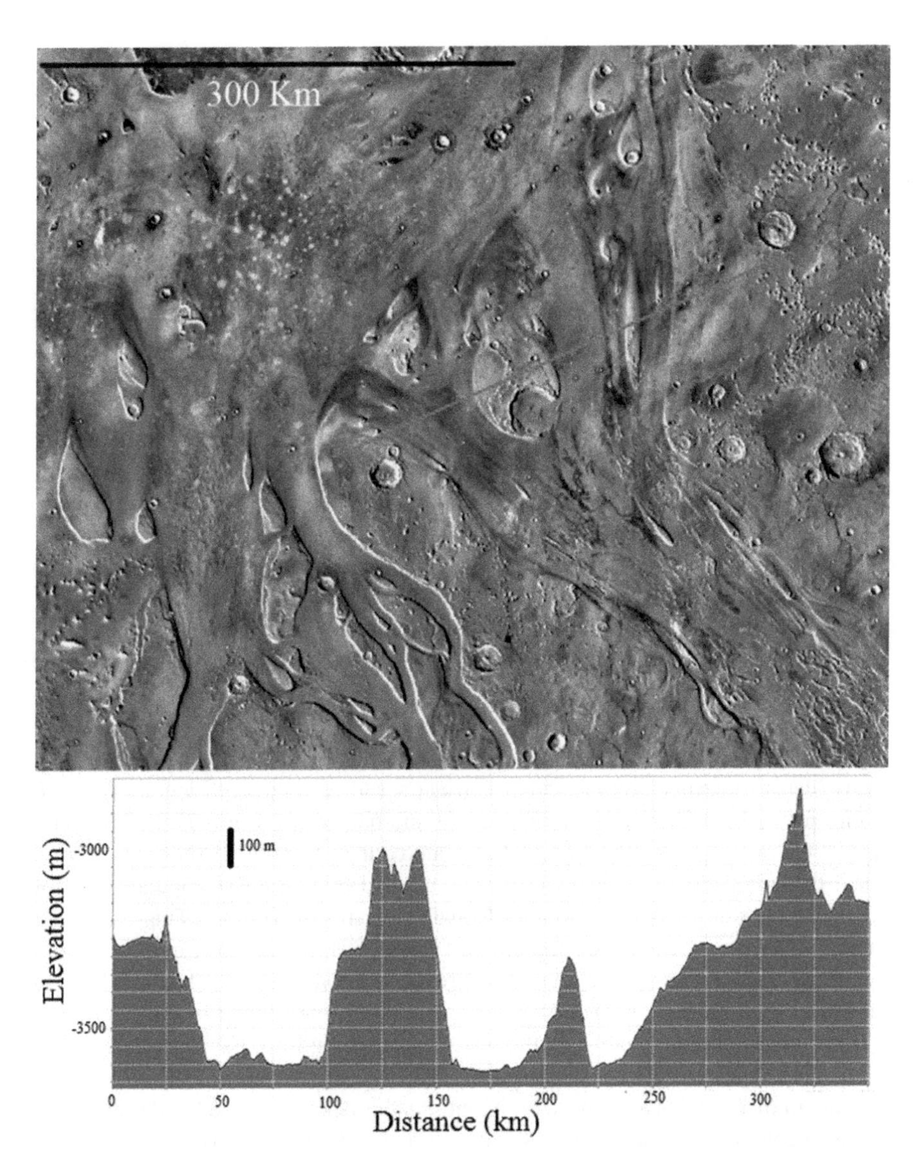

Fig. 4.32 Above: Ares Vallis (see its geographical position in Fig. 4.23) ends in the northern plains with canals and islands shaped like drops, as a consequence of erosion by ancient streams. Below: The elevation through the line in the figure shows the depth of the gorges. Top: THEMIS (Mars Odyssey, NASA), bottom: MOLA profile (MGS, NASA)

Similar results have been suggested for other outflow channels. However, for Ares Vallis, which is particularly steep, a water depth of 100 m would give a velocity of 27 m/, or nearly 100 km/h[4].

[4]Komar, P. D. (1980). Modes of sediment transport in channelized water flows with ramifications to the erosion of the Martian outflow channels. *Icarus*, *42*(3), 317–329.

What about the sources of these colossal streams of water? To give an answer, we need to better understand the different geometries of the Earth and Martian waterways. On Earth, streams develop a complex pattern controlled by the morphology and hydraulic features of the territory, from precipitation, slope, and tectonics. For example, an extensive volcanic area tends to create torrents arranged in a radial pattern that departs from the summit of the volcano to lower altitude. In addition, the main rivers receive water from the confluence of smaller tributaries, these by torrents, and the latter, in turn, by ever smaller streams down a hierarchical scale. Thus, a main river has many branching tributaries, so that the whole river network resembles a tree.

On Mars, however, these levels of hierarchy do not exist: at most, we may find a main waterway with a few tributaries. There is therefore a substantial difference: the rivers that produced the erosion channels form a single main stream of water. The differences do not end there. Following the longitudinal profile of a typical terrestrial river (the elevation along the direction of the stream) and of a Martian water course, there are substantial differences. While terrestrial watercourses are concave upwards, Martian profiles tend to be more linear. All of this seems to indicate different water sources for the two planets. On Earth, precipitation is the main water source in the catchment basin. On Mars, the valleys feeding outflow channels often start at an amphitheater such as Ravi Vallis (Fig. 4.24) and the absence of large dendritic systems seems to indicate a source within the ground itself, such as a sudden melting of permafrost.

But recent surveys based on high resolution images (HiRISE) have shown that many Martian torrents resemble the terrestrial ones to an impressive degree. Warrego Valles in southern Thaumasia exhibits a high drainage density of 0.53 per kilometer (this is the ratio between the length of all watercourses from the main river down to all of the smallest streams in a given area, divided by that area), compared to, for example, a value of 0.3–1.2 for the Kohala volcano on Hawaii. This incredible drainage system occupies the southern part of the Thaumasia block in the southern hemisphere. The bifurcation ratio measures 2.28, compatible with that of many terrestrial hydrographic networks, which most likely means that when the valleys formed about 3 billion years ago, the water was coming from the atmosphere. This brings us to another Martian mystery (Fig. 4.33).

The northern plains are covered with enigmatic rocks, embraced under the name of the geological formation of Vastitas Borealis. It is the most geographically extensive formation on Mars, covering a high fraction of the entire planet. Vastitas Borealis coincides with the position of the hypothetical ocean, and according to some researchers, is precisely the remnant of that ancient ocean. In fact, this unit is made up of many different types of rock, formed by water, volcanoes, and wind transport (this topic, in part, superposes with the → Ninth Mystery; here, we shall be more focused on the nature of the Vastitas Borealis formation and its enigmatic morphologies, while the Ninth Mystery will deal with the ocean on Mars from a more general point of view).

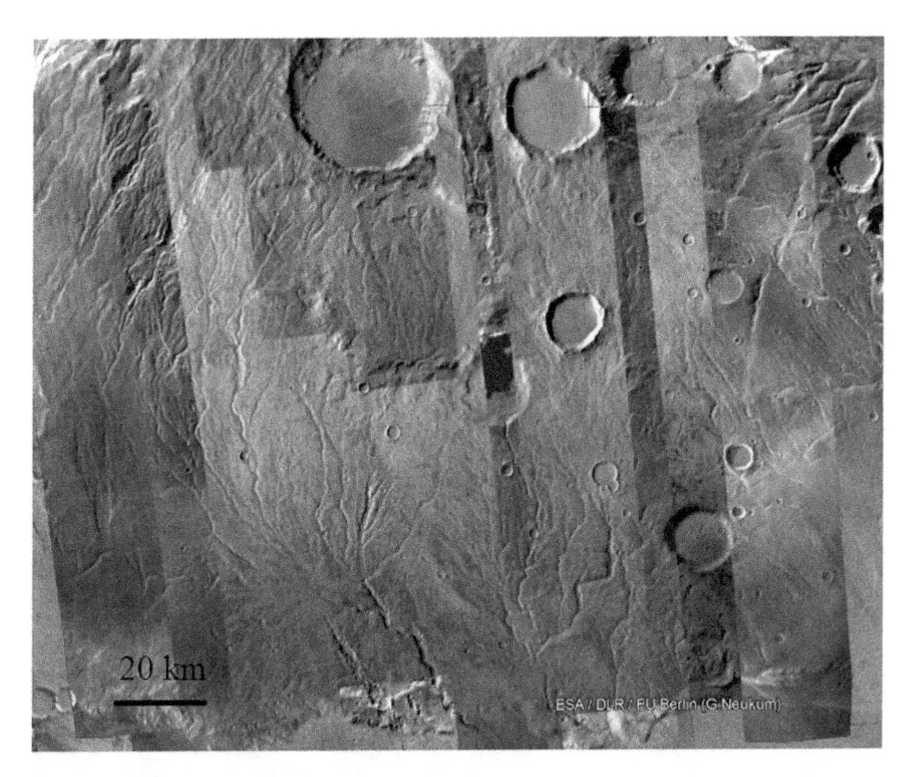

Fig. 4.33 River networks in Warrego Valles of the Thaumasia region. CTX mosaic, courtesy of Google Earth. CTX (MRO, NASA) mosaic on Google Mars platform

To appreciate both the thickness and uniformity of this formation at once, refer to Fig. 4.34. It again shows the two largest impact basins on Mars, which we already know well: Utopia and Hellas Planitiae. Similar in size (Utopia is a bit wider), and likely forged at approximately the same time: 4–4.1 billion years ago. Yet their elevation profiles are quite different. On the one hand, Utopia is flat, only 1 km deep, nearly indiscernible from the flatness of the northern lowlands. Hellas, on the other hand, is 8 km deep and is one of the very first Martian features that can be noticed at a glance. Thus, the original bottom of Utopia was at least 7 km deeper than the present floor. Why this difference? The apparent difference between the two basins is that Utopia is in the northern lowlands, Hellas in the southern uplands. Sediment and lavas must have filled Utopia to a much greater degree than Hellas. Not everything inside Utopia is sediment, however. Lava flows from Elysium Mons straight into the Hellas basin are evident, especially in the infrared (Fig. 4.35 shows a detail of the USGS geological map based on such images). These lava flows are more recent and occurred after the sedimentation on Utopia ended.

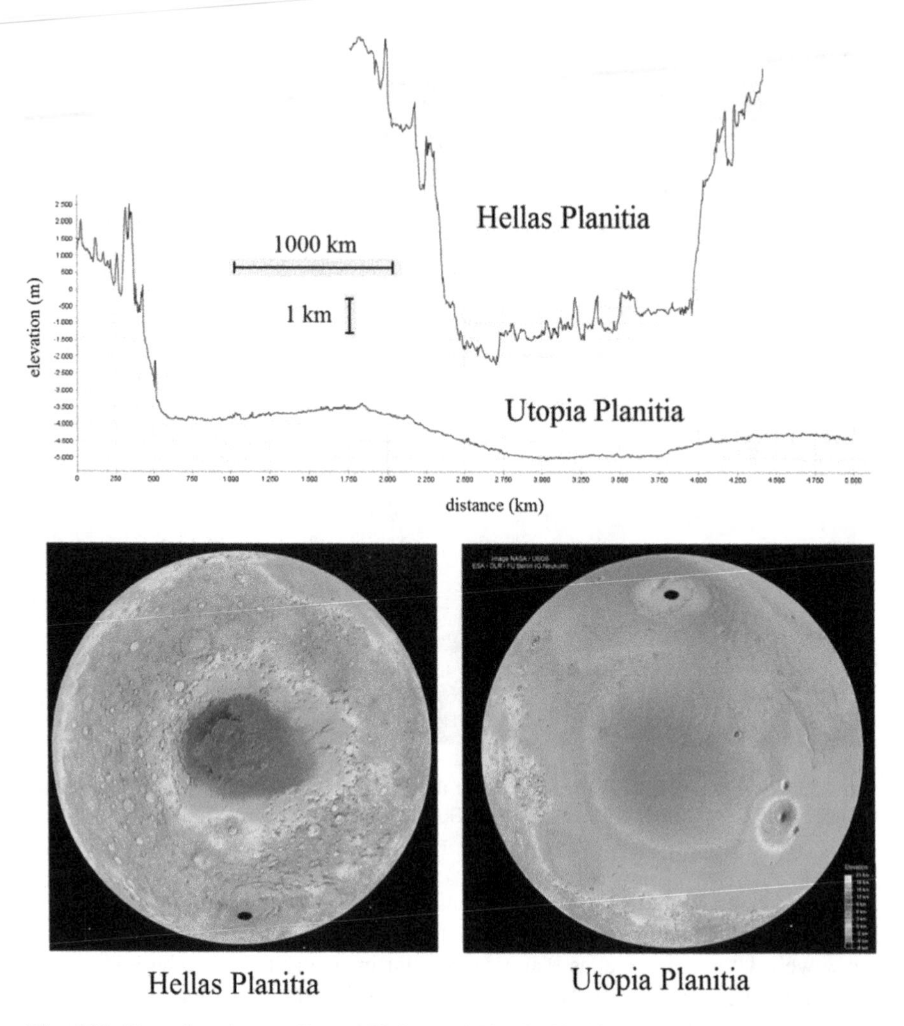

Fig. 4.34 Top: elevation profiles of Hellas and Utopia Planitiae (top, Hellas profile shifted vertically by 6000 m). Bottom: colorized MOLA map centered at the Hellas and Utopia Planitiae. Utopia is the circular area in the center. Southwest of Utopia is the Isidis impact basin. To the east, the Elysium Mons volcano. MOLA (MGS, NASA)

Other indications are based on image inspection at a smaller scale. Figure 4.36 shows cracks in the soil of Adamas Labyrinthus in southern Utopia Planitia. Such morphologies may form in different ways. Are they the product of icy soil, like the ones in Fig. 4.2, or perhaps contraction–expansion cracks physically corresponding to the ones formed in concrete, or more similar to water-driven mud cracks, like the ones in Fig. 4.31?

Polygonal soils in permafrost do not grow into the giant ones like those in Fig. 4.36. However, similar polygons of the right size are known to exist in the

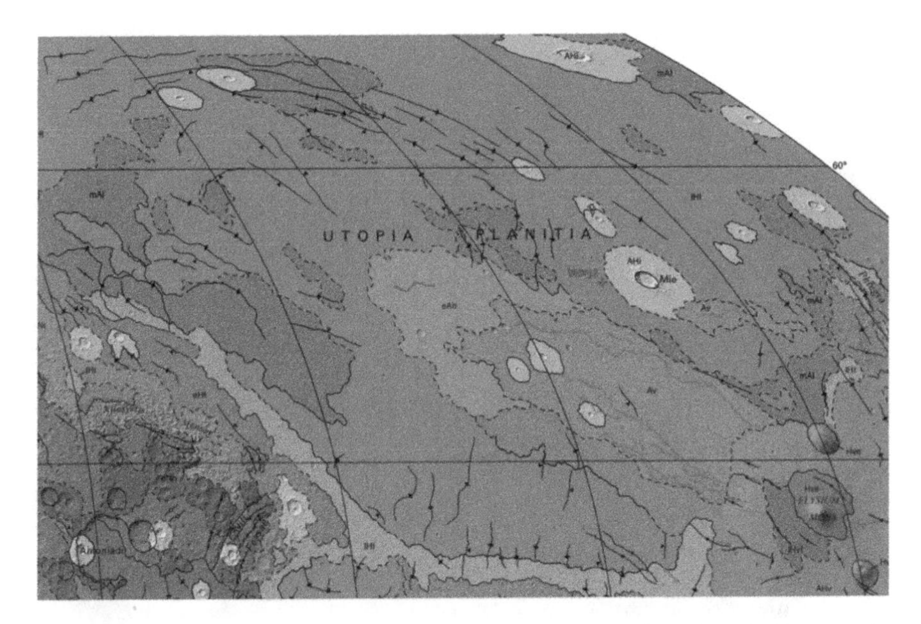

Fig. 4.35 A detail of the USGS geological map showing the lava flows from Elysium into the Utopia Planitia impact basin. Enigmatic ridges are shown as black lines with a dot. Part of USGS geological map

subaqueous environment,[5] and thus, by definition, they do not require desiccation conditions to form. Quite the opposite: they are present only in waters deeper than 500 m and are typically observed as seismic images, like in the Vøring plateau offshore Norway. Thus, this possible explanation not only indicates but actually requires an entirely subaqueous environment. There is no obvious mechanism of deep-water polygons in the terrestrial ocean, but it seems to involve, in some way, the rheological and geochemical behavior of oceanic clays. Figure 4.36 on the right shows a more detailed image of cracks. Note also the muddy appearance of the impact craters, which postdate the cracks.

NINTH MYSTERY: Was There an Ocean on Mars?

If we dried up the ocean water, the Earth would expose the flat ocean bottom at a depth of approximately 5000 m below sea level (Fig. 4.37); in marked difference with the ocean, continents exhibit average altitudes slightly greater than the sea level. Earth thus reveals a dichotomy, oceans as opposed to continents, reflected in the distribution of terrestrial elevations. In Fig. 2.15 (left), the distinct peak at −5000 m corresponds to the negative elevation of the oceanic bottom, while the other peak at +300 m signals the altitudes of the land. Drawing a similar figure for Mars, we find a rather similar result (Fig. 2.15, right), with less distinct peaks. The peak at high

[5]Moscardelli, L., Dooley, T., Dunlap, D., Jackson, M., & Wood, L. (2012). Deep-water polygonal fault systems as terrestrial analogs for large-scale Martian polygonal terrains. *GSA Today*, *22*(8), 4–9.

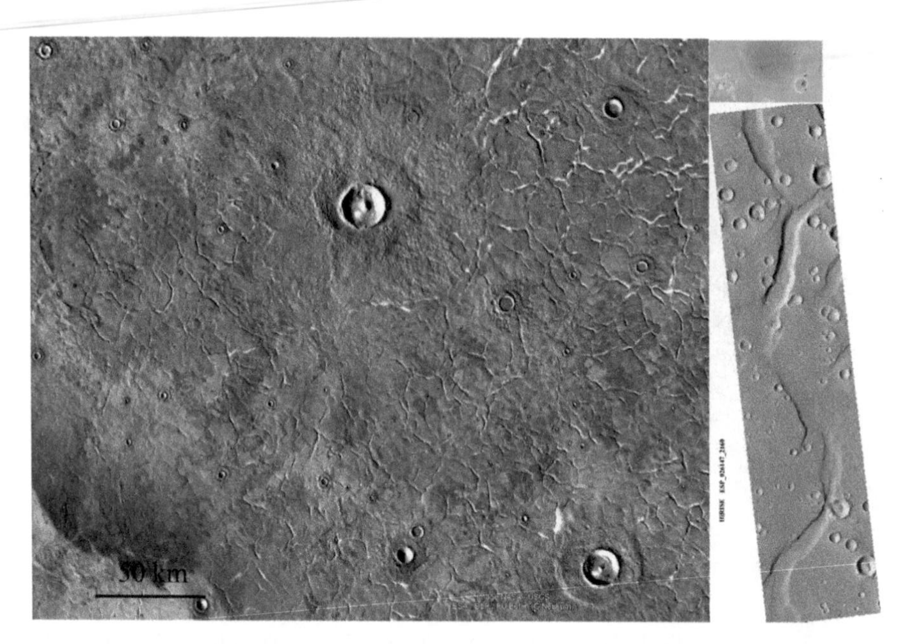

Fig. 4.36 Sopra: Enigmatic polygons in Adamas Labyrinthus inside Utopia Planitia. The cracks are similar to the desiccation patterns on Earth, but much larger in scale. Notice that some polygons meet at 120°, others at 90°. Image THEMIS. Right: detailed HiRISE image of one of the cracks in Adamas Labyrinthus. The image is about 5 km wide. Left: THEMIS (Mars Odyssey, NASA); right: HiRISE (MRO, NASA)

elevation (about 1200 to 1300 m above datum) now corresponds to the southern uplands of Mars, and the one at about −4000 m identifies the northern lowlands. The resemblance between the hypsographic curves for Earth and Mars is intriguing, but the similarities do not end there.

In blatant contrast with the mountainous and rugged surface of the southern uplands, the northern lowlands are smooth, as if sedimentation had been pervasive and uniform (Eighth Mystery). Although lava flows from the volcanoes surrounding the lowlands (from Tharsis, Elysium and Syrtis Major) have contributed to filling the basin, we also know from the analysis of our own oceans and lakes that smooth and ubiquitous sedimentation is more typical of subaqueous, rather than subaerial, environments. There are at least three reasons for this, all of them a result of the water's competence in transporting fine sediment uniformly and across a long distance. Firstly, the buoyancy and drag forces exerted by ambient water on small particles affect the movement of the particles, compared to a free fall on land. That is, a pinch of sand dropped into water falls slowly and distributes more uniformly compared to a fall through air. Secondly, the same drag forces dramatically limit the independent movement of fine sediments; as a consequence, sediments follow the flow of water and may be dragged across long distances. This is the reason why water currents in the ocean are capable of transporting particles over hundreds or thousands of kilometers away from the source. In addition, in the terrestrial oceans, processes such as contour currents, waves and deep-water currents continuously stir

Fig. 4.37 Schematic image of the Earth if the oceans were dried up. Compare with a MOLA map, such as the one in Fig. 4.34. On both Earth and Mars, a marked difference (dichotomy) appears between the low, relatively smooth areas (corresponding to the oceans on Earth and the upper highlands on Mars) and the elevated, rugged continents, which, on Mars, would correspond to the southern highlands. Image 134581863 FOTOLIA/Jacek Fulawka

the waters, permitting re-circulation of fine sediment. While subaerial clastic sediments (e.g., fluvial) tend to be dumped *en masse* in correspondence with topographic hiatuses during which water loses competence, the subaqueous ones (oceanic and lacustrine) are more dispersed in the whole basin of deposition and susceptible to water currents, as well as being strongly affected by their granulometry. Thus, the tremendously thick deposit inside Utopia Planitia indicates an oceanic, rather than subaerial, sedimentation (Eighth Mystery). Thirdly, water has a mobilizing effect not only on small sedimentary and independent particles, but also on more competent subaqueous debris flows and landslides, which may acquire extraordinary mobility because of the lubricating effect of water. Another piece of information also has potential interest: there is a gradual northward decrease in elevation of the Vastitas Borealis formation with the distance from the dichotomy line. This also indicates oceanic-like control of the sediment input from the southern lowlands.

The geological map in Fig. 4.35 also shows some black lines parallel to the contour of Utopia Planitia. They are considered to be wrinkle ridges resulting from contraction. These ridges appear in many locations of the Vastitas Borealis formation, not only encircling Utopia Planitia, but also around Alba Mons. Notice that they are parallel to the contour lines, rather than being directed along the slope, and this is also true for the ridges parallel to Alba Mons. One possible interpretation is that such ridges are furrows produced by underwater currents. Contour currents are the result of gravity and the Coriolis force induced by the Earth's rotation. Contour currents develop parallel to contour lines, which is possible only if an ocean is present. Such currents may also erode the flanks of subaqueous slopes. Because the rotation period of Mars is only slightly greater than that of Earth, the Coriolis acceleration has comparable magnitude. Can the furrows along the slopes of Alba Mons and Utopia Planitia be the product of contour currents[6]?

What was the aspect of Mars during its oceanic youth? To answer this question, one needs to "fill up" the northern lowlands to the supposed level, assuming that a relatively small degree of uplift or downlift has taken place since that time. Digital elevation models thus give us a clue as to the extension and shape of the Martian ocean. Figure 4.38 shows the possible outline of the ocean, assuming an oceanic level in the northern highlands of −2000 m below datum. In the hypothesis that all of the significant volcanic features we observe today were already there, the ocean would have been arrowhead-shaped, with the arrow tip in Chrise Planitia. Notice, however, that Elysium Mons is between 2.5 Ga to 3.6 Ga old, and so it may have altered the area after the demise of the ocean.

Global dichotomy and the analogy with the terrestrial ocean floor encouraged the quest for possible signatures of the Oceanus. In a dried-up Earth like the one in the figure, we could easily track the coastlines. Even if unaware of the watery past of the Earth, we would notice the evident difference in elevation between continents and oceans. One could also spot shore morphologies created at the ocean-land boundary. In the same way, Mars researchers have initially looked for putative shorelines, with several possibilities having been suggested.[7] One of the best examples is a line known as the Deuteronilus level. It runs several thousand kilometers along Mars' dichotomy boundary, at an elevation of approximately −3800 m to −4000 m. Other presumed levels bear the names of Acidalia, Arabia, and Ismenius and have different outlines. One problem with these levels is that they are not at constant elevation, as one could expect for genuine shorelines. Although it has been argued that even terrestrial land-sea boundaries are often not perfectly level, usually, for our planet, such discrepancies can be traced back to tectonic or morphologic uplift of some portions in the involved areas.

Another problem can be appreciated considering our own planet once more. We know that coastlines do not exactly mark the end of the continental shelves (which

[6]De Blasio, F. V. (2014). Possible erosion marks of bottom oceanic currents in the northern lowlands of Mars. *Planetary and Space Science*, *93*, 10–21.

[7]Parker, T. J., Gorsline, D. S., Saunders, R. S., Pieri, D. C., & Schneeberger, D. M. (1993). Coastal geomorphology of the Martian northern plains. *Journal of Geophysical Research: Planets*, *98*(E6), 11061–11078.

Fig. 4.38 Mars Express image elaborated to simulate the presence of the Oceanus Borealis at a level −2000 m below datum. Courtesy ESA. ESA, mdf

are the "true" boundaries of the continents, visible as sharp cliffs in Fig. 4.37), but are more landward. Moreover, the level of the terrestrial oceans has not been constant throughout the geological times. In the Mesozoic, the era of the dinosaurs, the sea level was approximately 100 to 150 m higher than today. It seems a small figure, but translated into horizontal distances, it turns out that, in those times, the shore was something between tens to hundreds of kilometers landward. This change has occurred in a mere 100 million years, or just one tenth the possible duration of the Martian ocean. Considering that the Earth is climatically more stable than Mars,

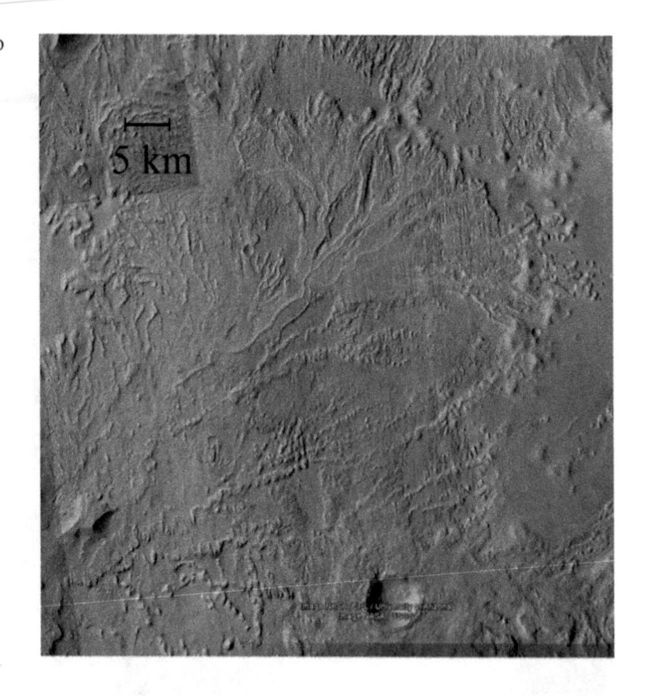

Fig. 4.39 A delta similar to those of Earth seems to indicate a region where water flowed as a river. Eberswalde crater in Margaritifer Terra, CTX mosaic. CTX (MRO, NASA) mosaic on Google Mars platform

it follows that putative Martian shores, even if they existed, might never be found unambiguously.

Other possible features indicating the demise of the past ocean can be investigated. Terrestrial rivers, when conveying their waters to the sea at the shore, tend to deposit huge volumes of sediment in the form of wide deltas. Thus, in retrospect, the presence of deltas may indicate the past position of a shore, from which the outline of an ocean can, in principle, be determined. If Martian deltas do indicate past locations where rivers delivered their sediment load to the ancient ocean, they should be approximately at the same elevation corresponding to the level of the shore.[8] Figure 4.39 shows a well-known delta on Mars inside the Eberswalde crater. It turns out that these deltas are relatively common at the dichotomy boundary, and that they tend to occur at a rather constant elevation: between −2700 m and −2300 m. Some of the investigated deltas face the putative boundary of the ocean, thus confirming the past existence of a common level for them, even though most deltas occur on the bottom of impact craters, which may complicate the interpretation.

We know from studies on our own planet that several morphologies are peculiar to the ocean. Identifying similar features in the northern lowlands would imply a further indication in favor of the past Oceanus Borealis. Figure 4.40 shows a fan-channel association on the Olympus Mons W aureole (Sixth Mystery). Note the roughly 90-km-long channel terminating in a fan on the basal level of Amazonis

[8]Di Achille, G., & Hynek, B. M. (2010). Ancient ocean on Mars supported by global distribution of deltas and valleys. *Nature Geoscience*, *3*(7), 459–463.

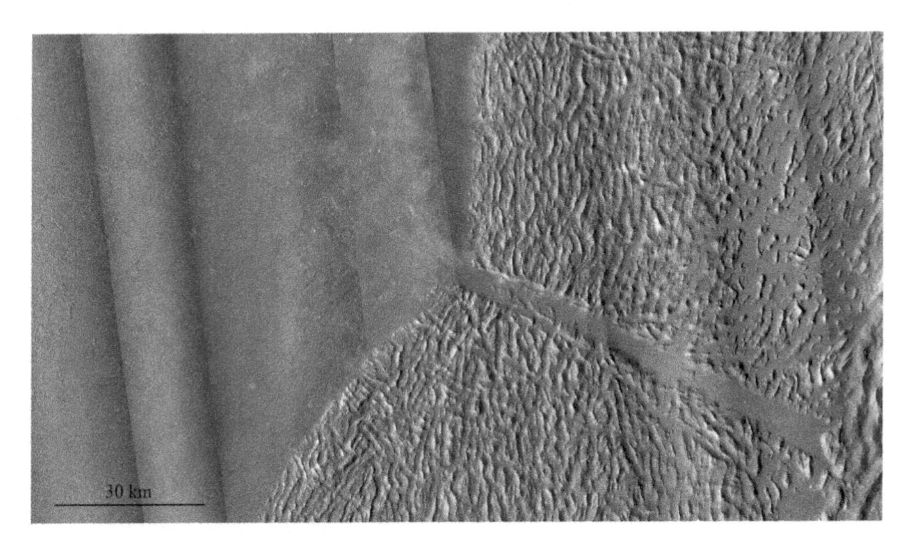

Fig. 4.40 A channel on the W aureole. Is it a submarine channel similar to the terrestrial ones? CTX (MRO, NASA) on JMARS platform

Planitia. The fan slopes very gently toward the plains of Amazonis Planitia, with a slope angle of about 0.35°, hardly discernible from the horizontal. Such angles are more typical of subaqueous, rather than subaerial, fans.

Other evidences of a past ocean on Mars include the presence of wide flow channels in correspondence with the boundary between the Noachian–Hesperian and the northern Amazonian terrains, the abrupt termination of valley networks at the alleged shoreline. Also, geophysical methods lend support to the idea that the northern lowlands were once filled with water. It has been found that the dielectric soil properties in the northern lowlands may indicate a massive amount of ice and the presence of phyllosilicates, minerals that typically form in water.

Addenda

- The atmosphere of Mars is extremely dry. The entire atmosphere spread on the surface would give a layer only one hundredth of mm thin.
- The water flows that sometimes develop from impact craters show that the surface of Mars is filled with long fractures, often arranged in frameworks.
- Some of the outflow channels are named after Mars in different modern or extinct languages: Kasei in Japanese, Ma'adim in Hebrew, Mawrth in Welsh, and Shalbatana in Accadic.
- Oceanic processes are effective in transporting sediments by means of down-slope processes like turbidity currents, submarine landslides, and debris flows. Do they explain the uniformity of the Vastitas Borealis formation?

Chapter 5
Atmosphere, Climate and Life on Mars

On the previous pages, we have examined many of the morphologies on the surface of Mars. But how can we explain river deltas, flow channels, and torrents on a planet that is today completely arid? The climate of Mars must have changed over the billions of years, from one that was warm and dense to the cool, thin, and arid atmosphere we see today. How did this happen, and why?

We know that the climate at the surface of the terrestrial planets is determined by the presence or absence of an atmosphere, the thin gas layer crucial to ensuring a stable climate on a planet. In the pages that follow, we will thus try first to understand the current atmosphere of Mars. We will see how, despite having a very low pressure, the Martian atmosphere is active. It gives rise to strange tornadoes several kilometers high and strong winds, or even storms that perturb the entire surface of the planet. In order to explain why Mars was a lot richer in water, we will discuss how Mars' atmosphere and climate may have changed.

Finally, the most important issue: life on Mars. Today, the climate of Mars and the absence of a magnetic field would not seem to accommodate life on its surface. But the highest atmospheric density and the amount of water available could have sustained life at some point in Mars' history. Are there any clues to fossils or bacterial life forms in the past on the Red Planet? Or any life forms still living today?

F. V. De Blasio, *Mysteries of Mars*, Springer Praxis Books,
https://doi.org/10.1007/978-3-319-74784-2_5

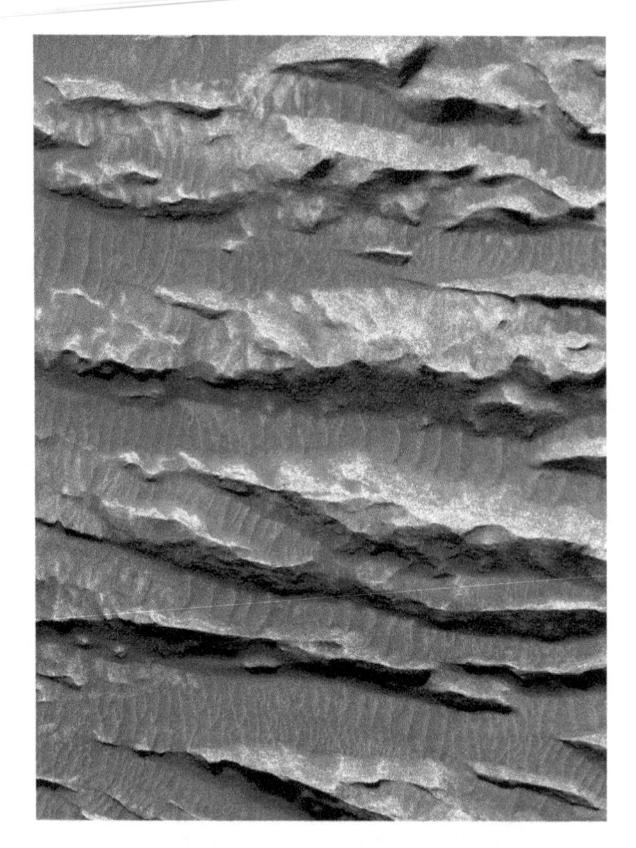

The above image shows a portion of a chaos area east of Valles Marineris (Arsinoes Chaos, −7° 02; 332.18° E), with parallel funnels excavated by windblown sand, a particular erosion form known as yardang. The bottom has then been re-filled by transverse Aeolian ridges (kind of like sand dunes in a desert, but smaller) parallel to the yardangs. Image HiRISE ESP_039563_1730 about 2 km across. North points to the right of the image. HiRISE (MRO, NASA)

5.1 The Atmosphere of Mars

Composition

Even before the space age, telescopic observations of Mars were fruitful for the study of the atmosphere, more than for that of the surface of the world, in which misinterpretations were numerous. The atmosphere has often given rise to gigantic and well-observed phenomena, even with the telescopes of the eighteenth and nineteenth centuries. Father Angelo Secchi noticed clouds in 1858; the polar ice caps, already known at the time of Huygens (1659), require the presence of an atmosphere to exist. We have also seen that the Nix Olympica, later identified as the great volcano Olympus Mons, had been noticed by Earth due to the clouds that may appear on its summit, similar to the Earth's orographic clouds.

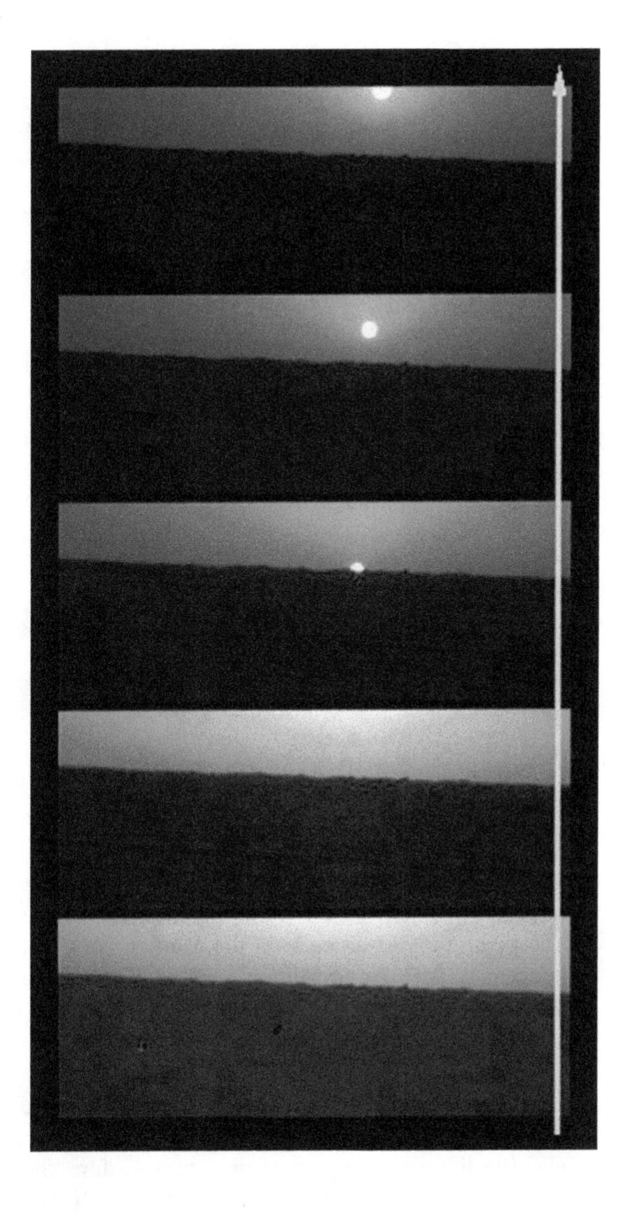

Fig. 5.1 Due to the suspended powder, the Martian dawn is preceded by intense brightness. The bottom-up sequence was taken from the Pathfinder rover Sojourner. Arizona State University, NASA. Image Pathfinder (Arizona State University, NASA)

Until the early 1960s, it was thought that the atmosphere of Mars was nitrogen, the gas that composes 78% of our atmosphere. One of the disappointments of the Mariner 4 was precisely the discovery of a thin atmosphere, certainly not comparable to our own. The ground pressure of about 600 pascals equals 1/500 of the Earth's. To find such low air pressure on our planet, we would have to rise to a height of 30 km. The idea that the atmosphere was thick was thus proven wrong. Although tiny, the air was sufficiently thick to keep material dust suspended by the Martian winds; and with it, also the hopes of finding life on Mars. Because of the suspended dust, Martian sunrises and sunsets are spectacular. Figure 5.1 shows the sequence of a dawn taken by the

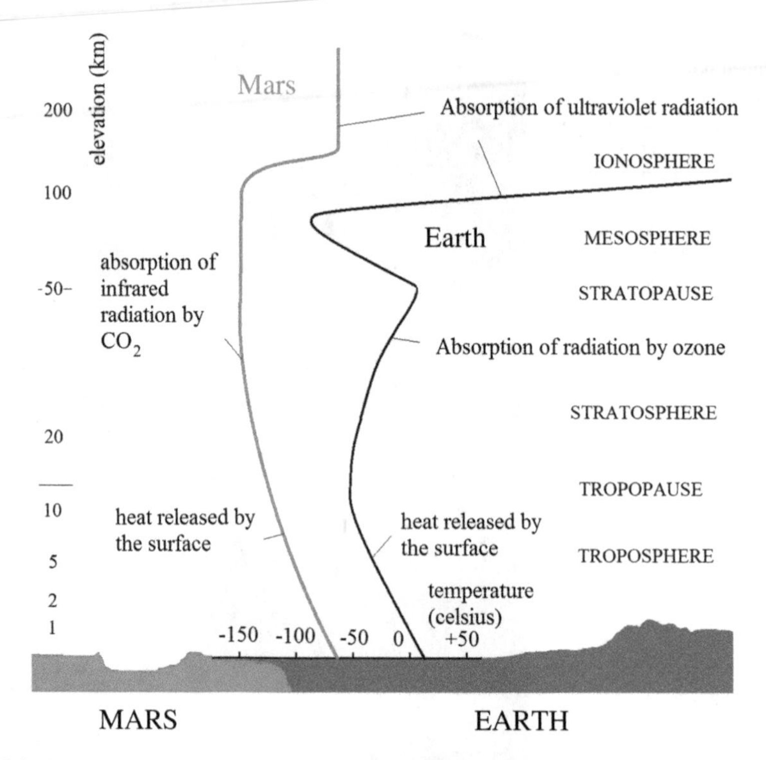

Fig. 5.2 Temperature of the atmospheres of Mars and the Earth as a function of the altitude from the ground. For the different atmospheric zones, the main mechanisms that heat the air are indicated. FVDB

Pathfinder. When the Sun is still below the horizon, the luminosity resulting from the light scattered by suspended particles almost looks like a second sun.

The percentage of nitrogen in the Martian atmosphere is only 2.7%; the main component is 95% carbon dioxide, CO_2. There are traces of argon, oxygen in molecular form, and carbon monoxide. The percentage of water in the atmosphere of Mars is very small, less than a tenth of Earth's. Despite this, cloud formation systems are observed, especially in high areas.

Both on Earth and Mars, the heat from the soil warmed by the Sun is transferred to the overlying layers of atmosphere. That is why the air temperature decreases with height. But this decrease occurs only up to a few kilometers, after which a new phenomenon ensues (Fig. 5.2). On the oxygen-rich Earth, ultraviolet sunlight splits some of the oxygen molecules into their two constituent atoms. These can either re-bind together or form a bond between three, instead of two, oxygen atoms. So, ozone is formed, a very efficient gas in absorbing ultraviolet light. For this reason, above 10–15 km, the temperature of the Earth's atmosphere begins to increase, reaches 0° at 50 km in height, and then decreases again. Not so on Mars, due to the different atmospheric composition. On Mars, the temperature decreases constantly up to 100 km in height, mitigated a little by the absorption of infrared radiation by the carbon dioxide molecules. The temperature goes down much more than on Earth, reaching −150 °C.

But also, on Mars, the highest layers of the atmosphere are heated by ultraviolet sunlight. This is the so-called thermosphere (also called an ionosphere on Earth due to the intense ionization state), where the temperature of Mars's atmosphere rises up to −70 °C, comparable to that at the surface.

The Martian Winds

Winds on Earth are created by temperature and pressure contrasts in air masses, with the addition of the Coriolis acceleration due to the rotation of the planet and the perturbation of the mountains or other topographic features. So it is on Mars as well, with some important differences. The temperature contrast between day and night is much greater on Mars than on Earth. The cold night air at the surface of the Red Planet tends to descend to the valleys, expanding in the Martian depressions. During the day, warm, less dense air rises to higher altitudes. In addition to these local phenomena, there is an average circulation around the planet, which is affected by its rotation. On Earth, the hot equatorial air tends to rise to tens of kilometers. The air reaches the stratosphere, where the temperature gradient is reversed. Thus, air masses cannot go upwards any further, and are forced to move horizontally. The air descends at about 30° latitude, forming what is called the Hadley cell. That is why the latitudes at 30° enjoy, on average, high pressure, with associated good weather. Likewise, at latitudes around 60°, there is a low-pressure zone, and the poles have high atmospheric pressure again at 90° latitude. This subdivision of the atmospheric circulation cells in steps of 30° latitude explains the prevailing movement of the air masses. At the equator, air moves from east to west (generating the so-called "trade winds"), while at latitudes of 30°, they move in the opposite direction. There is no stratosphere on Mars, so Hadley's cell manages to encase the air mass of the entire planet. The prevailing winds therefore stir the atmosphere from east to west in equatorial regions, and from west to east at high latitudes.

Another effect exacerbated by the thin Martian atmosphere is called the "thermal tide." As Mars rotates around its axis, warmth from the Sun behaves like a heat wave traveling along the surface from west to east. This phenomenon also occurs on Earth (with similar time frames), with the difference that our atmosphere is denser and does not respond immediately to the change in the Sun's warmth. On Mars, however, the change is immediate and generates winds that change direction during the day. A weathervane would rotate 360° during a Martian day.

The most outstanding atmospheric phenomena on Mars are generated when the planet is at the perihelion. During that period, due to the eccentric orbit, Mars receives the maximum thermal energy from the Sun. Sandstorms thus develop from the southern hemisphere (the one facing the Sun at perihelion). For unclear reasons, once every 3 years, such storms may be extremely strong and mobilize the whole atmosphere, hence the name Global Scale Storms (Fig. 5.3). The strong winds that develop, and especially the airborne particles, may be extremely hazardous for future exploration, as they have been for the rovers currently on Mars.

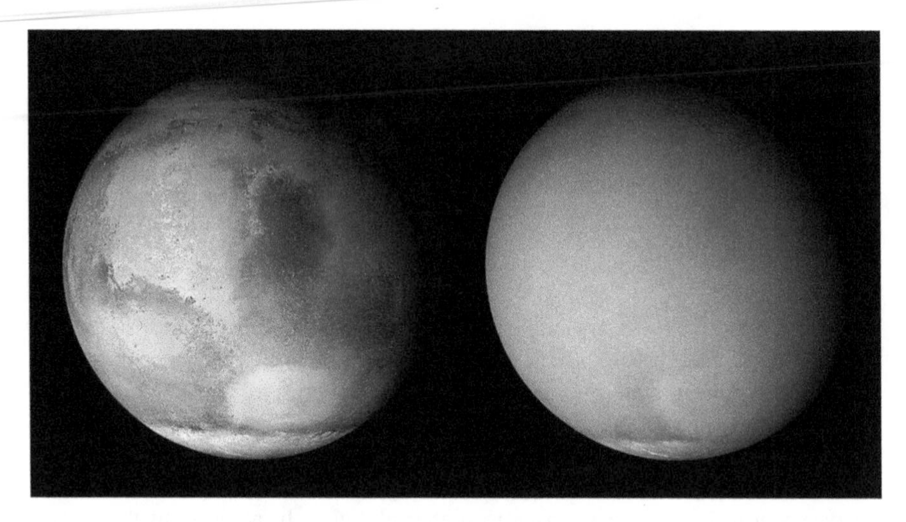

Fig. 5.3 Images taken at different times (June 2001 and July 2001) of tranquil Mars and of Mars exposed to a global scale storm. Note in the right image the haze developing during such events. The large elliptical feature in the south is Hellas Planitia. Images MOC (Mars Orbital Camera) on board MGS. Courtesy of Malin Space Science Systems and the California Institute of Technology. MOC (Malin Space Science Systems, MGS), NASA

Morphologies Created by the Wind

On Earth, the wind is more a meteorological than morphological phenomenon. Only in deserts does the absence of vegetation and water allow sand particles (usually quartz, the most resistant mineral) to be suspended. Sand gathers in sand dunes and mega-dunes of various shapes. The abrasive particles accelerating in the strong wind are capable of eroding the rock. Particles smaller than 1 mm are raised by the wind of the desert, making partly ballistic trajectories. Granules less than one tenth of a millimeter in size may remain in suspension for long periods of time. On Mars, the absence of water over the entire planet makes the non-cohesive regolith prey to the Martian winds, much more than on Earth. In fact, on Mars, Aeolic morphologies such as dunes, barchans, and yardangs abound in perfect forms. There are, however, differences compared to Earth. The quartz of our deserts, a light mineral, is replaced by the heavier basalt, abundant on the surface of Mars. But that is not the real difference. Have we ever wondered where the sand of the desert comes from and where it settles at the end? In geologically fast times, sand is transported far to non-desertic areas, where it ends up in sediments or is transported to the ocean. However, on Mars, granules do not have this chance. At least in the current climate devoid of water, and thus of cohesive sedimentation, granules travel incessantly across the surface of the planet, accumulating in barchans and dunes that are then removed, and so on, in never-ending cycles.

The dunes often accumulate in boreal regions like those in Fig. 5.4, covered by an icy mantle. The wind sculpts the sand of the largest dunes, drawing an arc with the convex part upwind. These large accumulations migrate slowly, responding to the

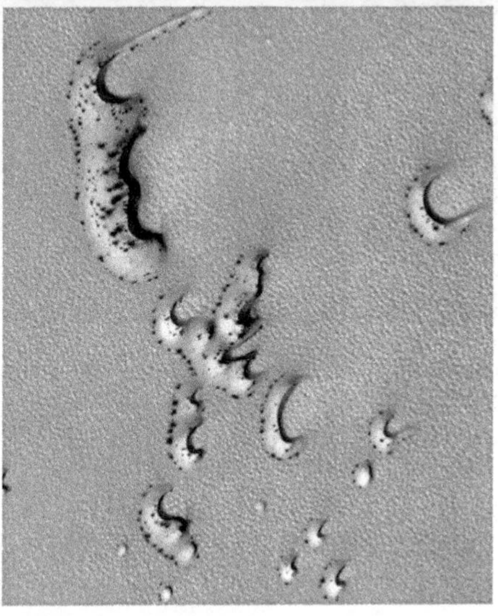

Fig. 5.4 Top: sand dunes in a terrestrial desert. Bottom: dunes on Mars in the boreal regions. HIRISE ESP_016177_2575_RED. Image width about 5 km. Coordinates 77.4° N, 126.0° E. Top image 146442024 FOTOLIA/Guy Bryant, bottom: HiRISE (MRO, NASA)

movement of single grains. At the low Martian atmospheric density, wind can reach a speed of 100 km/h, carrying grains more easily than on Earth. Launched against a rock, such a fast particle disintegrates, and perhaps that is why sand-size granules are rarer than we should expect. Just to avoid the impact of these bullets—small but fatal—the Viking landers had metal masks to protect some critical parts in the event of a storm. But the rocks of the planet have no protection, especially if loose. So, yardangs are created, a particular form of erosion also known on Earth (Fig. 5.5). Terrestrial yardangs appear as eroded rock columns, especially at the base (where sand granules are more abundant), or as small channels separated by tens of meters. Those of Mars are enormous, and funnel-shaped rather than columnar. They can be a few kilometers wide and hundreds of kilometers long (Fig. 5.5). It is amazing to think that they have been incised by the wind.

If you were on Mars during a sandstorm, you would try to shelter yourself behind a crater where the wind shear is smaller. That is why on the lee side, the craters often show a "comet" (Fig. 5.6). Within the crater and in the downwind areas, the weaker wind allows smaller particles to accumulate. Because the basalt sand-size particles

Fig. 5.5 Left: Aeolic erosion forms known as yardangs in Turkey. Martian yardangs (right) may be hundreds of kilometers long. Image HRSC H9373_0000_ND3. Coordinate circa 5° N, 212° E. Large scene about 80 km wide, magnified image at the bottom about 8 km wide. Left: image 45831131 FOTOLIA/sunsinger, right: HRSC (Mars Express, ESA/DLR/FU)

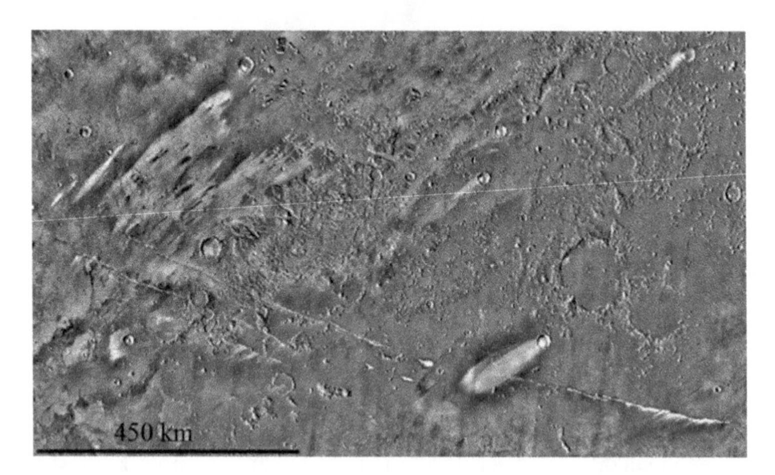

Fig. 5.6 The swaths resulting from the Martian wind blowing against craters and fractures are evident in this daytime infrared THEMIS image. White areas have a higher daytime emission and a lower thermal inertia; they are therefore composed of finer particles than the dark areas. THEMIS (Mars Odyssey, NASA)

are dark, while the smaller powders are light red, winds generate albedo variations around obstacles. Since the smaller particles have a lower thermal inertia, thermal inertia is also affected.

Technical Box 8: Thermal Inertia Measurements

It is difficult to withstand the tremendous heat emanating from the beach sand in the early afternoon Sun, and walking is possible only on the shadows cast by umbrellas. After sunset, however, the sand becomes cool, and it is enjoyable to dive your feet into on summer evenings. Contrastingly, a large boulder never becomes too hot, even under the strongest sun, while at night, it keeps the heat much longer than sand. This different behavior has to do with the conductivity of sand compared to rock. Being composed of small granules, the sand conducts the heat much more slowly than the non-fragmented rock. So, on a sunny day, the heat of the Sun cannot penetrate deeply; the sand warms enormously on the surface, while the deepest parts remain cold. But the heat penetrates better inward into a boulder, heating it more uniformly. At night-time, the heat accumulated in the sand is only superficial and goes away quickly, while it takes a lot more time to dissipate the much more penetrating heat inside of the boulder, so that the boulder stays warm, even at night. The property of maintaining heat for a long time is known as thermal inertia. How to use this data for Mars?

A hot body emits much energy in the infrared, a frequency band between 0.7 and 3 microns. The THEMIS (Thermal Emission Imaging System) camera onboard the Mars Odyssey spacecraft spent 15 years in orbit around the Red Planet between 2001 and 2016. The camera is sensitive to both visible and infrared light. If the difference between day–night temperatures is large, the material on the surface must have a low thermal inertia, and thus consists of sand or dust. A more moderate temperature difference between day and night implies high thermal inertia and the presence of rocky material on surfaces. Therefore, by measuring the temperature difference between day and night, it is possible to distinguish fine granular material from rock. Figure 3.17 shows how such data is used to infer the presence of both dust and compact rock on Mars in Tithonium Chasma. The information gained in this way, however, regards only the superficial veneer, while it is useless for deep layers. The use of infrared light reflection, in addition to that in the optical canal, also allows for analysis of additional spectral bands, improving mineralogical analysis from remote sensing. Thus, in addition to thermal inertia, infrared analysis allows us to infer mineralogy.

Dust Devils

Tornadoes on Earth are produced by the effect of strong winds on wide and flat areas. Mechanically, a tornado is a stable structure, because within these vortexes, the high air velocity creates a pressure low that acts as suction for the surrounding air. More modest in terms of size and destructive capacity are the dust devils, frequent in the deserts (Fig. 5.7, left). On Mars, very high dust devils have been observed, like the one captured by HiRISE and shown in Fig. 5.8. They usually occur during the

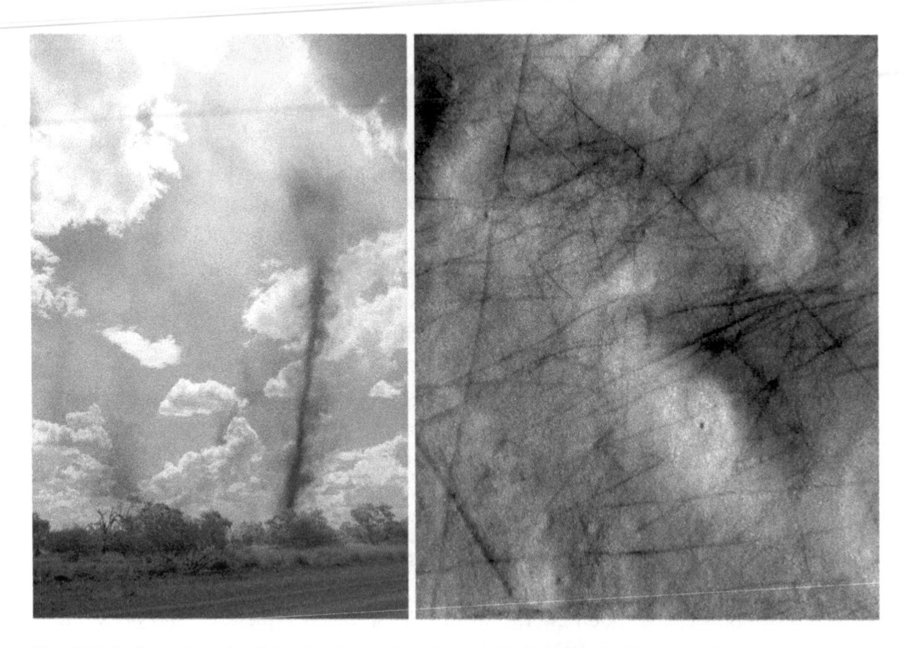

Fig. 5.7 Left: a dust devil in the Australian desert. Terrestrial devils are much shorter than the Martian ones. Right: Black stripes due to the passage of a dust devil in the Amazonis Planitia plain. Image HiRISE PSP 008870_2370. 55° N, 180° E. NASA/Lunar and Planetary laboratory-Univ. of Arizona. Left: image 127488734, FOTOLIA/totayla, right: HiRISE (MRO, NASA)

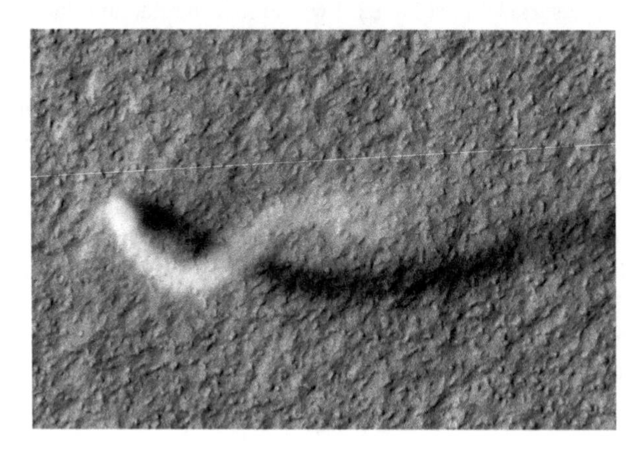

Fig. 5.8 A dust devil runs in the plains of Amazonis Planitia. The base of the devil is 30 m long. Image HiRISE ESP_026051_2160 taken in February 2012 at 35.8° N, 207° E. Courtesy NASA/JPL_Caltech/University of Arizona. HiRISE (MRO, NASA)

afternoon, when the temperature differences generated by the warm Sun are at their maximum, and preferably in the southern part of the planet, in the period between spring and autumn. One common factor in the images below is the product created by these devils. When a dust devil passes through a sandy area, it removes the lighter dust cover, unveiling the darker underlying soil, and so leaving streaks at its passage

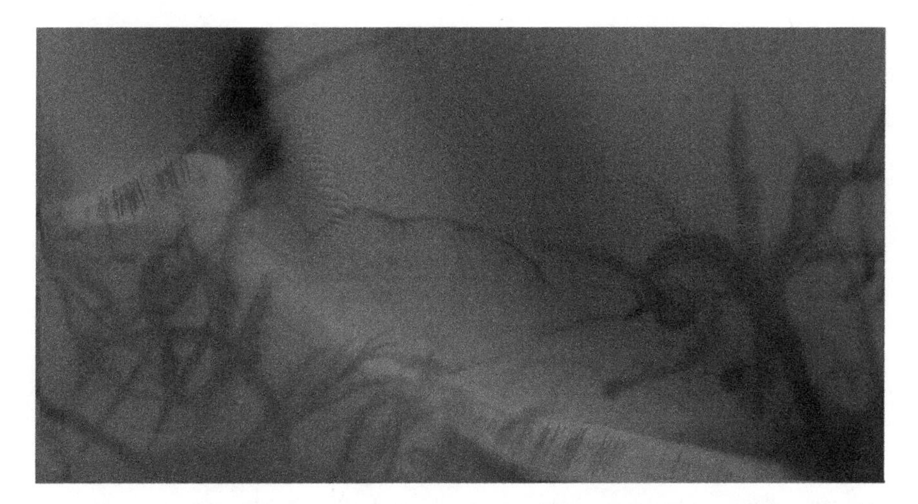

Fig. 5.9 Dust devils do not simply move linearly, but, responding to the local wind, temperature, and topography, they often develop arcuate paths. Image HiRISE ESP_014426_2070. Horizontal width about 2 km. 26.659°, 62.810° E. HiRISE (MRO, NASA)

(Fig. 5.7, right). The lines that are formed often have curved geometric motifs and appear more like a form of abstract art than a form of wind erosion (Fig. 5.9).

Present-day Changes on Mars

In a clear contrast to the previous chapters describing Martian morphologies that have been preserved for billions of years (lava flows millions of years old are considered very young!), Mars' meteorology shrinks time to a human dimension, showing us a still active planet. Comparing two images taken at intervals of some years, sometimes a slight difference pops up, for example, in the pattern created by some dust devils. Since it is difficult for major changes to occur, new morphologies found in high-resolution images are never more than a few kilometers across. Wind is not the only geological force documented on the contemporarily-changing Mars. Figure 5.10 shows a dusty avalanche at the north pole of Mars falling from a height of about 700 m, consisting of a probable combination of ice water, rock and ice of CO_2. These are probably seasonal phenomena, triggered by the sublimation of CO_2 from the polar caps. Other phenomena occurring between two consecutive observations are small superficial landslides and impact craters. An observation that brings together both superficial landslides and impact cratering is shown in Fig. 5.11. The vast superficial dark landslide was not there before the acquisition date (March 31, 2010), and the small crater enlarged in the image on the right was also absent. The hypothesis is that the fall of a small meteorite excavated the crater, causing the avalanche of superficial regolith at the same time. Other changes in contemporary Mars are the recurrent slope lineae considered in Chap. 3.

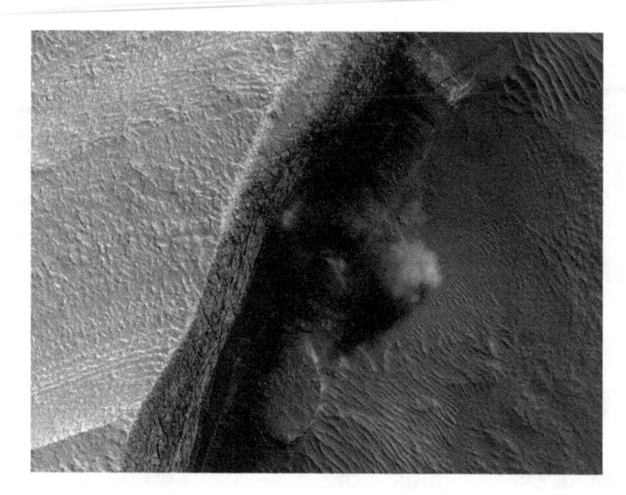

Fig. 5.10 HiRISE caught this avalanche in the making on January 27, 2010, at the northern pole of Mars. Image HiRISE ESP_016423_2640. HiRISE (MRO, NASA)

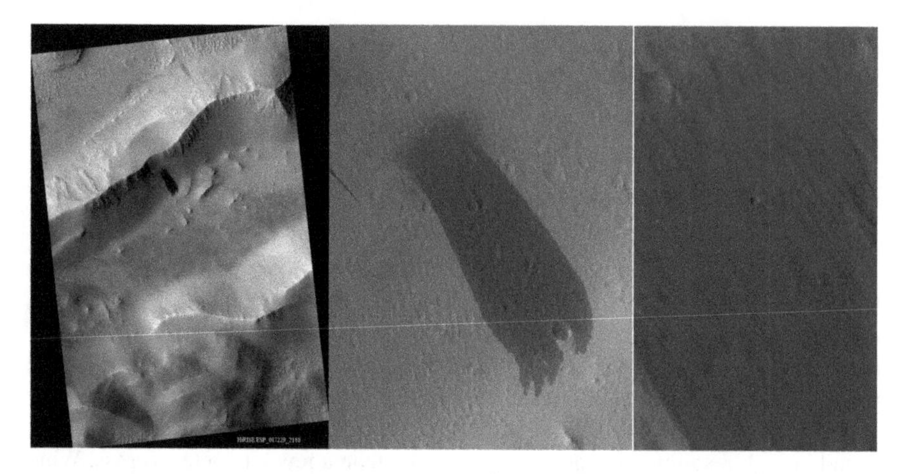

Fig. 5.11 A superficial landslide in the aureole of Olympus Mons, probably due to the fall of a meteorite (the crater is magnified on the right). HiRISE image ESP_0172229_2110, explanation by Alfred McEwen. Left image is 5 km across. HiRISE (MRO, NASA)

Other enigmatic structures are believed to have been excavated by dry ice (frozen carbon dioxide, the main component of the Martian atmosphere), and not by water. Fig. 5.12 illustrates the strange few kilometer-long channels directed to the inside of the Russell crater, in the southern part of Mars. They appear to be flows of water-rich detritus, frequent on Earth. However, the channels end abruptly, a difficult thing to explain given the streams of water-rich debris, which would end at the slope break with a delta-like deposition. The sudden disappearance of the canals suggested to the researchers that blocks of dry ice sliding downslope were

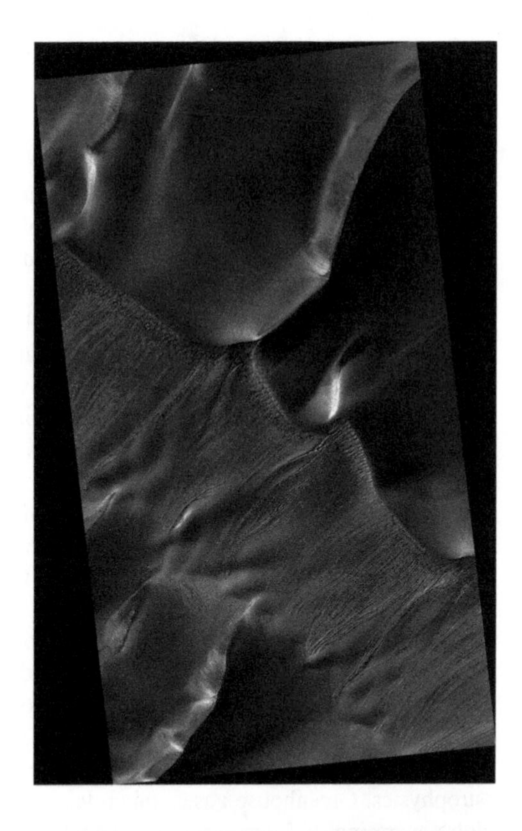

Fig. 5.12 Channels attributed to blocks of dry ice that slide along the sand dunes. HiRISE PSP image 001440_1255; coordinates: −54.2°, 12.9°E. HiRISE (MRO, NASA)

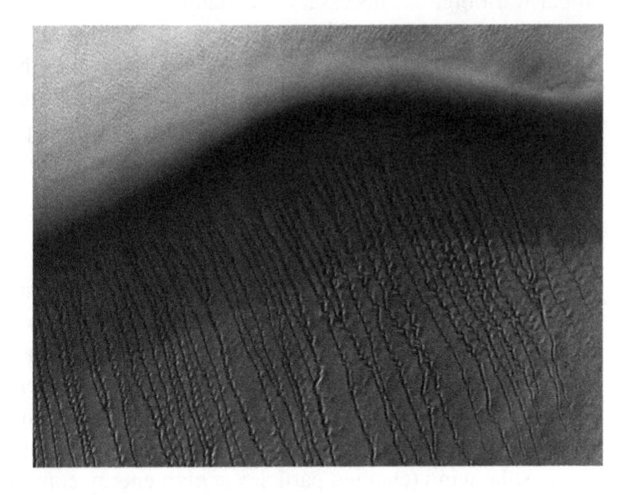

Fig. 5.13 These channels sometimes give rise to meandering pathways, probably incompatible with the ice block hypothesis. Coordinates −45.30°, 67.21° E. HiRISE (MRO, NASA)

involved. After complete consumption, the blocks have sublimated, returning to the atmosphere, which would explain their sudden disappearance. However, it is difficult to explain the meandering patterns observed in the image in Fig. 5.13 as the product of sliding dry ice, even though the style of erosion appears to be exactly the same.

5.2 Mars Climate and Life

A Radical Climate Change

Despite the dryness of present-day Mars, this book has shown that morphologies created by running water are common on the planet. Some have involved the melting of permafrost following meteorite impact, and can be explained by the intense heat emanated by the impact. Conversely, there are also cases of more widespread channels excavated by running water, as well as ancient river networks apparently similar to the terrestrial ones, probably indicating precipitation over the vast areas. There are also indications, albeit more controversial, of standing bodies of water in lakes and in the immense Oceanus Borealis, perhaps one billion years or more after the formation of Mars. Thus, all of this information indicates that ancient Mars must have been much wetter than today.

At present, there is quite a lot of ice in the Martian soil. The problem of explaining the presence of water on Mars in the past is turning this ice into water: wetter actually means warmer. What could have increased the temperatures at that time? Could Mars have received more energy from the Sun, for example? The theory of stellar evolution tells us that after a main sequence star like the Sun is born, its temperature slowly increases with time. Astrophysical models, confirmed by observations in our Galaxy, indicate that two billion years after its formation, the Sun was 30% fainter than today. This exacerbates the problem of a wetter Mars rather than solving it. We thus need to invoke mechanisms internal to the planet—climatology rather than astrophysics. Greenhouse gases, much feared today in view of alleged problems of global warming, but necessary for maintaining the present temperatures above the freezing point of water, could increase the temperature substantially. The most significant greenhouse gas is CO_2, but the problem is that, in the present day, this gas is rare on Mars. Though it constitutes the bulk of the atmosphere, it does so in very tiny amounts, while the CO_2 layer at the poles is presently too thin. This gas thus appears to be totally insufficient to account for a significant temperature increase. What about rocks? The recent discovery of carbonates in the Martian rocks covered by lakes appears interesting. However, carbonates are so rare on Mars that they could be sold as jewelry.

Could CO_2 have escaped into space? The smaller gravity field and radius at the Martian surface signifies that the escape velocity of molecules is about one half that on Earth, and thus Mars may have lost a significant amount of its gases, including CO_2. Mechanisms that could increase the loss rate are very complex and involve the electric charging of molecules through the sputtering mechanism, occurring when molecules are ionized in the upper atmosphere. The absence of a magnetic field allows solar wind (charged particles at high energy coming from the sun) to hit such molecules hard, dispersing them in space. It should also be mentioned that ice could act as a protective medium, and thus isolate water. So, the presence of ice may also help explain the presence of water without too dramatic an increase in temperature.

However, no serious attempt to model the variations in the Martian climate can be attempted without accounting for yet another effect. Planets change their orbital characteristics with time, as we know very well from our own planet. The tilt angle (obliquity) of the Earth's axis with respect to the ecliptic and also its eccentricity have changed over time, giving origin to cycles on the order of about 26,000 and 100,000 years, respectively. These variations may partly explain the temperature variations at the Earth's surface, as recorded in the sedimentary rocks and in the ice caps, even though the effects are not fully understood, and have probably led to only very small variations in temperature. This is because our Moon provides rotational stability to the Earth, and thus orbital variations are limited. Mars, which lacks massive moons, has experienced far more dramatic amplitude variation of the obliquity and eccentricity of the Martian orbit. Calculations show that there have been periods when the axis was inclined at nearly 90°. Under conditions of larger obliquity, it is believed that the polar caps may have grown to lower latitudes (Fig. 5.14).

Where did the water come from and where did it go afterwards? The bombardment of comets and chondritic meteorites containing a lot of water during the phase of planetary accretion may explain an ocean several kilometers deep, but there are major uncertainties as to the transport rate of water by celestial bodies and of retention of water on the surface. The outgassing of water from the mantle probably formed an atmosphere whose lighter molecules of hydrogen escaped into space, energized by the energetic ultraviolet photons of the young Sun. This destiny was followed by light noble gases, but also by CO_2, which was pushed outward by the solar wind. It is believed that 3.5 Gy ago, the reservoir of water on Mars still

Fig. 5.14 Artist's conception of Mars in a situation of high obliquity. Image 161172300 FOTOLIA/dottedyeti

contained a quantity of water equivalent to 35–115 m spread all over the surface of the planet. However, the very late (Amazonian) age of exposed terrains in the Northern Lowlands interpreted as oceanic sediments may indicate that the Oceanus Borealis persisted to this very late age. The climatology of Mars and further discussion about the demise of water on Mars will be one of the major problems for planetary scientists to solve.

TENTH MYSTERY: Is There or Has There Been Life on Mars?

In 1984, a group of researchers picked up a set of meteorites at Allan Hills, in the southern area of Victoria Land in Antarctica. Antarctica is a dust bin for waste from space. Meteorites fall everywhere in the world, but it is difficult to distinguish a banal local stone in a garden from a meteorite, and not everyone is as lucky as the British gentleman who witnessed a fall into his own home garden in England (doubly fortunate, we could say). In Antarctica, where the geological layers are covered by a kilometer-thick sheet of ice, any rock visible through and above the ice cannot be terrestrial. Thus, the sample classified as ALH 84001 was collected and classified generically as achondrite, a group of stony meteorites derived from the asteroid belt. But the meteorite turned out to be far more interesting than expected.

The real story of the ALH84001 meteorite began much earlier. Four billion, one hundred million years ago, when the first crust on Mars had already consolidated, lava flowed slowly and cooled on the Martian surface. Subsequently, at a time impossible to determine but on the order of 1.5 Ga ago, carbonates were deposited in the rock fractures (possibly caused by the fall of a nearby meteorite) carried by an aqueous fluid. The rock, so contaminated remained there for another period of at least a billion and a half years. About sixteen million years ago, the fall of a meteorite cast part of this rock into space. The debris remained in space for sixteen million years, halfway between the orbit of Mars and Earth, until about 13,000 years ago, when, as Earth was beginning to emerge from the last glaciation, the meteorite fell on the Antarctic glacier.

ALH84001 is not the only meteorite that has come from Mars. Most of the meteorites from Mars are generically called SNC meteorites (Shergottites, Nakhlites, Chassignites, designated by the names of the locations where they were collected for the first time). Notice that some meteorites, including ALH84001, do not fit into any of the three classes of SNC. Altogether, over one hundred SNC meteorites are known, representing several points of impact on Mars. The Shergottites, the most numerous, derive from a special basalt, and show an age of 180 million years, probably that of their ejection from the Red Planet. The Nakhlites and Chassignites have a composition like that of dunite, a rock made up of olivine. They came from the bottom of a magma chamber, where the heaviest minerals settled. Their age of 1.3 Gyr indicates not only that meteorites do not come from the asteroid belt, but also that magmatic differentiation processes were active on Mars at that time.

ALH84001 became famous 12 years after the discovery at Allan Hills. In 1996, some researchers noted strange forms stretched within the carbonate spheres (Fig. 5.15, right). They were in the same form as terrestrial bacteria, though much smaller (less than 100 nm, at least double the size of mycoplasmas, the smallest

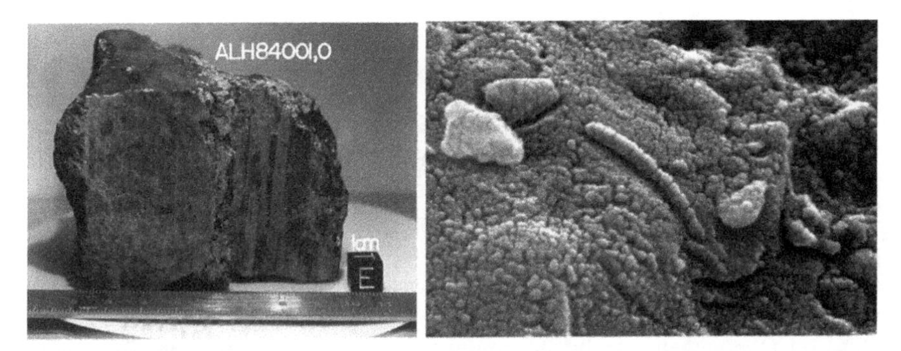

Fig. 5.15 The meteorite from Mars ALH84001 and an SEM image of its alleged fossil bacteria. NASA/JSC/Stanford University

known bacterial forms.) The minimum 200 nm limit seems essential for earth life, since ribosomes, made of multiple RNA and protein molecules, have a diameter of at least 20–30 nm. For this reason, many years after this discovery, there are many doubts as to whether these forms derive from life-related activities. Even the presence of magnetite on the meteorite, which some bacterial forms use for orientation in the Earth's magnetic field, is not considered a definitive test today.

Among the instruments on board the two Viking landers, there were two small biochemistry labs for measuring possible traces of life. The instruments were able to perform three experiments in complete autonomy. The first test was based on the idea that life always releases waste gas: carbon dioxide from animals and molecular oxygen from plants. After dampening one sample of scraped Martian soil, it was measured as to whether some of the microbes had released one of these gases. And in fact, something was measured: a rapid oxygen peak and a much slower release of carbon dioxide. While the latter could indicate some form of plant life in the Martian soil, the rapid oxygen peak was, in fact, too rapid, and a chemical reaction was suspected, rather than the presence of life. The Sun's ultraviolet light could have created oxygen-borne water on the ground, which, in contact with the water splashed by the micro-laboratory, broke in water and oxygen. So, this experiment did not lead to anything certain.

In a second experiment, an attempt was made to nourish hypothetical microbes with an amino acid broth. On Earth, we would observe the production of carbon dioxide in the biochemical activity of organisms. To distinguish the carbon dioxide released by microorganisms from that of the Martian atmosphere, part of the broth contained the carbon isotope 14, instead of the ubiquitous isotope 12. If the microorganism digestion products contained isotope 14, it meant that they had fed on the Viking broth. And indeed, this was observed: it appeared that some alien microbes had enjoyed the Earth's recipe. However, the results could still be interpreted as a purely chemical reaction involving hydrogen peroxide. The third experiment was intended to reveal carbon compounds—the essential bricks of life—but did not find anything. The final verdict from the three experiments was that Viking probably saw no signs of life. However, there have been recent claims that the experiments' outcomes could be more complex and perhaps a re-interpretation of data would be needed.

Nearly 40 years later, Curiosity took many photographs and analyses of the Martian soil in the areas surrounding its landing site. Although it did not find life directly, it did document the lacustrine conditions that could be favorable to life. Some researchers even took a step forward, interpreting certain morphologies photographed by Opportunity and Curiosity as stromatolite mats. Stromatolites are symbiotic associations between algae and bacteria, which, on Earth, form macroscopic columnar structures, detected in the fossil record since the Precambrian. However, there seems to be no consensus, nor outright rejection, from most researchers.[1]

A possible signature of life is methane, a gas that implies either geological or biological activity. Methane burps were detected in 2003, only to disappear a few years later. Ten years after that, Curiosity did find methane. Today, it is believed that the signal for methane was a consequence of contamination by the rover itself. The question of methane outbursts on Mars thus remains unsettled.

The rovers have also documented alleged Egyptian statues, metal artifacts, and monkey skulls (Fig. 5.16). How to explain these oddities? We may invoke Occam and his razor: the scientific principle that in the absence of any conclusive proof, the simplest explanation should be preferred. Is it easier to believe that the statue in Fig. 5.16 is the strange result of a game between rocks and shadows, creating the impression of an artifact (a well-known psychological effect known as pareidolia), or that billions of years ago, there was an ancient civilization on Mars, composed of beings like us?

Epilogue

Clearly, the mysteries listed in this book, and many more, are still a part of the agenda of planetary scientists, and so it would impossible for anyone to say that this is the last word. However, the reader might be disappointed if I, the author, did not at least partially disclose his opinion about at least some of these mysteries.

It must be evident from the previous pages that the writes is in favor of the hypothesis of a vast amount of standing water on Mars. I think that the evidence for both the existence of lakes on Mars and of the Oceanus Borealis is compelling. There is no way to transport sediment and smooth out such sediment over a vast area other

[1]For reinterpretations of the Viking experiments, as well as novel alleged indications of life on Mars, see, for example: Navarro-González, R., Vargas, E., de La Rosa, J., Raga, A. C., & McKay, C. P. (2010). Reanalysis of the Viking results suggests perchlorate and organics at midlatitudes on Mars. *Journal of Geophysical Research: Planets*, *115* (E12); Noffke, N. 2015. Spatial associations, and temporal succession in Terrestrial Microbialites. Astrobiology 15, 169-192. Bianciardi, G., Rizzo, V., & Cantasano, N. (2014). Opportunity Rover's image analysis: Microbialites on Mars?. *International journal of aeronautical and space sciences*, *15*(4), 419–433.

Fig. 5.16 Is it a game of shadows or a real figurine made by a humanoid civilization, as photographed by Opportunity (the "artifact" is enlarged at the top right)? NASA

than submarine processes. Sedimentation in Vastitas Borealis has been pervasive and has likely filled the lowlands with kilometers of sediment. As a corollary, the age of the northern lowlands should coincide with the age of the demise of the ocean. Most likely, both lakes and the Ocean survived well into the Hesperian period, but also into the beginning of the Amazonian. Moreover, I suspect that water was present when Olympus Mons was growing; otherwise, it is difficult to explain the gravitational instability that culminated with the aureole landslides.

I would say that Valles Marineris opened up "simply" as a consequence of the Tharsis bulge, which tore apart the Martian crust. The comparison with the road concrete swollen from below is explicit, albeit very simplistic. Then, ice in the Martian soil began to sublimate, loosening the soil. This can be explicitly seen in the catenae, i.e., the chains of collapse pits that adorn the surrounding areas of Valles Marineris. Weakened by this sublimation effect, landslides began, mobilizing hundreds of cubic kilometers of Martian soil at once. In my opinion, more exotic explanations are less likely. Finally, the number one mystery: the global dichotomy. External or internal? I personally prefer the internal hypotheses, but nobody knows, obviously, which of the many explanations is more viable.

However, at the end of this book, I cannot help but notice one intriguing thing. From a purely geometrical viewpoint, the problem of dichotomy would decrease if one were to imagine an originally uniform crust on a planet with smaller diameter. Then, one could imagine the planet expanding differentially, unveiling the mantle in the north, and so creating global dichotomy. The kinematic model of expansion that could give a reasonable match is similar to the opening of a flower, with the southern hemisphere being stretched and the northern plains created new, and then blanketed by sediments. Obviously, this is completely unproven and difficult to justify from a physical point of view. Theories of Earth expansion have continued to be discussed for one full century, even though the proponents have never managed to convince the scientific community, partly due to the difficulty in finding a mechanism for expansion. Another, more technical book will consider water on Mars and the global dichotomy in more detail.

Technical Box 9: Explore Mars Yourself

This short box is intended for students who have not (yet?) been captured by this fascinating field of science, or simply for people interested in planetary science who would like to live the excitement of planetary exploration.

Studying Mars at a research level requires training in one or more disciplines, like Earth Sciences, Physics, Engineering, or Chemistry. Normally, researchers are employed under some form of temporary or permanent contract at a scientific institution like a University or national research center. Mars and the planets are a democratic kind of science. Firstly, all of the data are available to the public, which is usually not the case for terrestrial field data. Secondly, Mars can be studied from one's own desk. The impossibility of directly smashing Martian rocks under an hammer holds just as much for the most important Marscientists as it holds for you. If you are interested in the study of Mars (or any of the other celestial bodies of the solar system) and even if you are not a professional, but rather an amateur, a student, or a scientist who does not specialize in planetary geology, you can still analyze the images of the main probes sent to Mars yourself, since they are in the public domain. Once you have examined many images and, even better, if you have a background in geology and geomorphology, you could really get to find something new, or reinterpret a morphology in an original way. Opening one of the images (even at random!) from one of the internet sites listed below, there is a significant chance that you won't be able to understand the origin of some of the features you observe. It is likely that some scientists have studied those features and come to a conclusion as to their origin. However,

(continued)

Technical Box 9 (continued)
more often that many people would believe, some morphologies have no obvious explanation, even to specialists. Thus, the choice of encountering some real Mars riddle by perusing a great number of images is certainly high.

The first thing you need to do to explore Mars yourself is to get acquainted with the optical and infrared images, because these images will be the easiest to examine without further treatment. Radar, nuclear, or reflectance techniques are somewhat more advanced and are not considered here. The most important images and data to consider are the following:

- Viking 1 and 2, both landers and orbiters. Although the Vikings have been the two missions with the highest impact on Martian studies, most of their images have been superseded by successive missions. However, many of them are still useful and a bunch of fundamental publications have been written based on Viking data.
- MOLA altimeter data (MOLA laser altimeter on board Mars Global Surveyor, MGS) are essential for morphologies that are at least a few kilometers long, while the resolution will not be sufficient for smaller structures.
- MOC (Mars Orbiter Camera), also aboard MGS, consisted of three different elements sensitive to different wavelengths. The black-and-white has provided images with a resolution of 1.4 m per pixel.
- HiRISE, aboard the Mars Reconnaissance Orbiter (MRO), is a high resolution camera providing extremely detailed pictures of Mars, showing details down to one third of a meter per pixel. Coupled to HiRISE is the context camera CTX.
- CTX has a much lower resolution than HiRISE (8 meters per pixel), but covers the whole of Mars.
- THEMIS (Thermal Emission Imaging System) are thermal (infrared) images obtained by the Mars Odyssey mission, which spent 15 years in orbit around Mars between 2001 and 2016. The camera is sensitive to both visible and infrared light. The use of infrared light reflection, in addition to that in the optical canal, extends TES analysis to additional spectral bands, improving mineralogical analysis from remote sensing. In addition to mineralogy (which, however, requires a non-obvious data treatment), infrared analysis allows us to tell a rock-covered area from a dust cover.
- HRSC is the high resolution camera on board the European Mars Express.
- The rovers Spirit and Opportunity, and the ongoing Curiosity.

Where can you retrieve the images? For reasons of space, here is a list of a small number of internet sites where images can be downloaded, without any pretense towards completeness. One useful web page is:

(continued)

Technical Box 9 (continued)

- http://themis.asu.edu/maps

which contains, on just one page, links to THEMIS, HiRISE, MOC, CTX, HRSC, and Viking. A map allows you to choose the area of interest and to point at the images you want to download. Try to be selective, or too many images will confound you. Another useful link is the following "MARS global data set" that allows for an overview of the whole planet:

- http://jmars.mars.asu.edu/maps/

More images from NASA and ESA can be downloaded from their respective web pages. A particularly friendly software is Google Earth, from which you can choose to be redirected to Mars. This software is also being used more and more for professional studies. Typically, you'll be able to navigate the whole planet by uploading the CTX images, but the other images can also be retrieved from within the program. Another recommended, user friendly software is jmars:

- https://jmars.asu.edu/

which will enable you to examine the surface of Mars (and other planets and moons, too) and draw MOLA elevation profiles and many other analyses.

- http://photojournal.jpl.nasa.gov/catalog

is the catalog of NASA photos.

- https://hirise.lpl.arizona.edu/

is the HiRISE catalog, which, for analysis, should be supplemented by the Hiview program, downloadable from

- https://www.uahirise.org/hiview/

HiRISE images examined with Hiview are so detailed that one can appreciate details that are as small as less than a meter. In practice, the examination of such images will take hours of time. You can also propose new targets to NASA through the Hiwish program.

Finally, yet another field of exploration is the one of Martian meteorites. The discovery of new sites for Shergottites, Nakhlites, and Chassignites, or perhaps a new class of Martian meteorite, is always possible.

Addenda

- At the equator, the Martian temperature swings between 0° in the middle of the day to −100° at nighttime.
- In the realistic movie "The Martian," a storm causes major harm to a human settlement on Mars. Due to the low density of the Martian atmosphere, wind shear on Mars (the strength exerted by the winds on objects) is relatively small. Although the strength in the movie appears exaggerated, we don't know if winds may grow stronger for a very active Sun (see figure below).
- Although it had long been suspected that some meteorites could originate from Mars, it was only by examining the gas content in the bubbles of ALH 79001 that scientists found the same composition and isotopic ratios of the Martian atmosphere.
- Due to its present low atmospheric pressure, on Mars, water in a glass would boil and evaporate away within minutes or freeze. How thus could water in rivers, lakes, and oceans survive in the past?
- In 1976, the Viking 1 orbiter imaged the Cydonia area of Mars. One photo showed a strange face that was looking up, a Martian sculpture tens of kilometers long. The myth of the face on Mars was thus born, and the old hypothesis of Martian civilizations revived. Viking's image was missing a pixel embedded in the nostril, increasing the visual impact of the face. As improved images from later missions have shown, there is no mysterious face in the monolith of Cydonia (see figure below).

Top: typical landscape in Jordan where parts of the movie "The Martian" were shot. Bottom: The "face on Mars" imaged by the Viking 1 orbiter; image 124594211 FOTOLIA/trypsilon, Viking 1 (NASA)

Appendix: Location of Images on Mars and Earth (t = top; b = bottom; Fr = frontispiece; r = right; l = left)

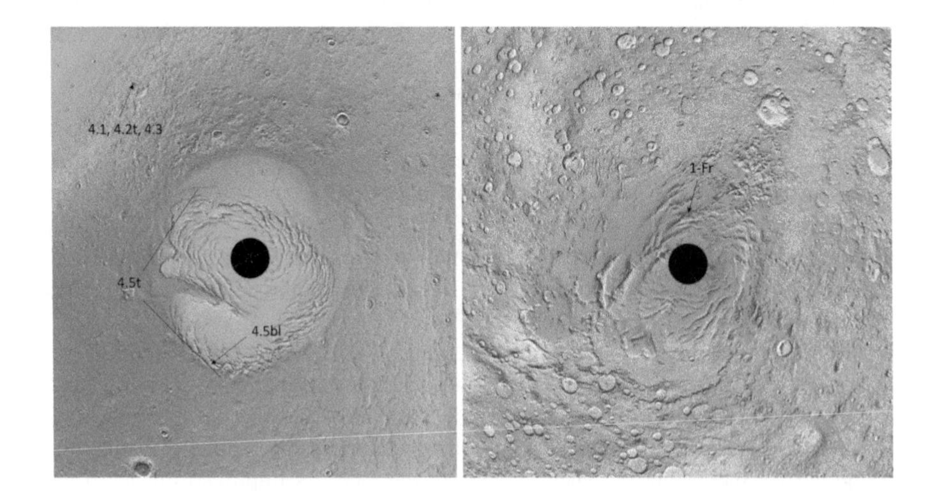

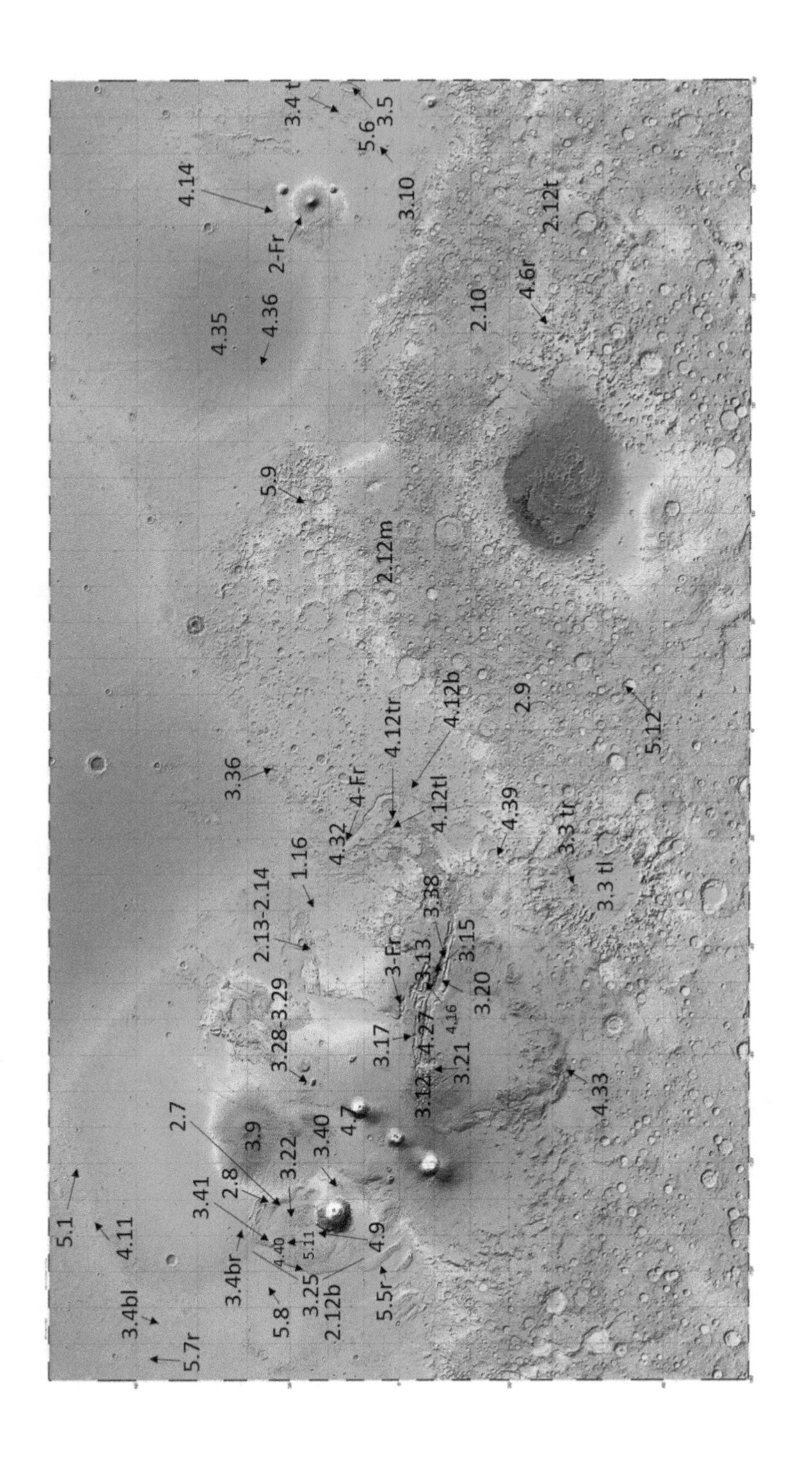
3.4 t
5.6
3.5
4.14
2-Fr
3.10
2.12t
4.35
4.36
2.10
4.6r
5.9
2.12m
2.9
5.12
4.12tr
4.12b
3.36
4-Fr
4.12tl
4.39
4.32
3.3 tr
1.16
3.3 tl
2.13-2.14
3.38
3-Fr
3.13
3.15
3.28-3.29
3.20
4.16
3.17
4.27
3.21
3.12
4.33
2.7
3.9
4.7
3.40
3.22
2.8
3.41
5.1
4.11
4.40
5.11
4.9
3.4br
3.25
2.12b
5.8
5.5r
3.4bl
5.7r

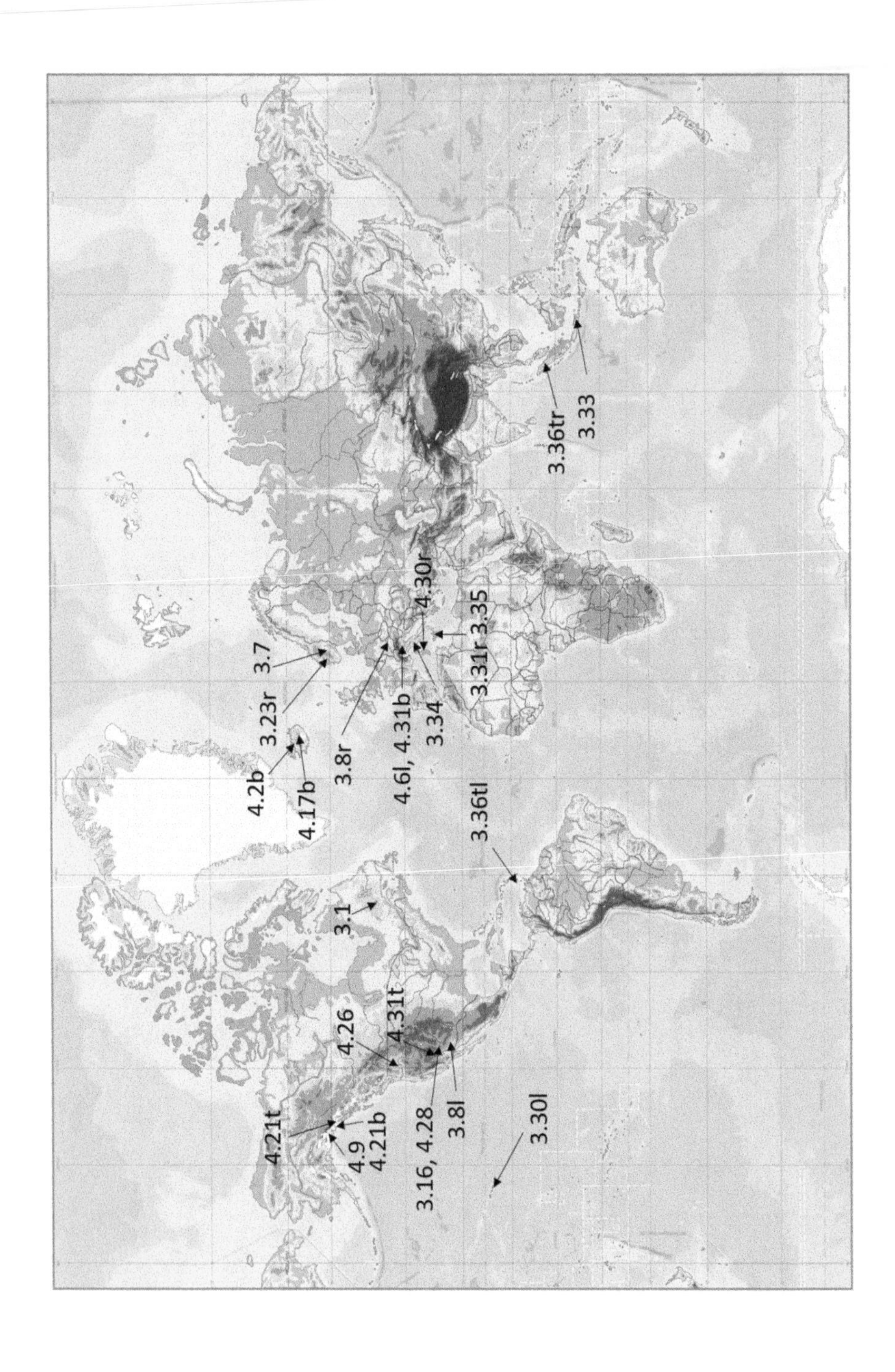
4.21t
4.9
4.21b
3.16, 4.28
3.8l
3.30l
4.26
4.31t
3.1
3.36tl
4.2b
4.17b
3.23r
3.7
3.8r
4.6l, 4.31b
3.34
4.30r
3.31r
3.35
3.36tr
3.33

References[2]

Aldrin, B. 2013. Mission to Mars. My vision for space exploration. National Geographical Society, Washington, USA. (*)
Baker, V.R. 1982. The channels of Mars. University of Texas press, Austin, USA. (***)
Barlow, N. 2008. Mars. An introduction to its interior, surface, and atmosphere. Cambridge University Press, Cambridge, UK. (***)
Bell, J. 2008. Mars 3-D. A rover's-Eye view of the red planet. Sterling, New York, USA. (*)
Cabrol, N.A., and Grin, E.A. (Eds.). 2010. Lakes on Mars. Elsevier, Amsterdam, The Netherlands. (***)
Carr, M.H. 1996. Water on Mars. Oxford University Press, New York, USA. (***)
Carr, M.H. 2006. The surface of Mars. Cambridge Univ. Press, Cambridge, UK. (***)
Chapman, M. (ed.). 2007. The Geology of Mars. Evidence from Earth-based analogs. Cambridge University Press, Cambridge, USA. (***)
Ezell, E.C., Ezell, L.N. 2009. On Mars. Exploration of the Red Planet 1958–1978. Dover Publications, Inc., New York, USA. (**)
Forget, F., Costard, F., Lognonné, P. 2008. Planet Mars. Story of another world. Springer Praxis, Berlin, Germany. (*)
Frankel, C. 2005. Worlds on fire. Volcanoes on the Moon, Mars, Venus and Io. Cambridge University Press, Cambridge, UK. (**)
Harland, D.M. 2005. Water and the search for life on Mars. Springer Praxis, Berlin, Germany. (**)
Hartmann, W.K. 2003. A traveler's guide to Mars. The mysterious landscapes of the red planet. Workman, New York, USA. (*)
Kargel. 2004. Mars. A warmer wetter planet. Springer Praxis, Chichester, UK. (**)
Kieffer, H.H., Jakosky, B.M., Snyder, C.W., and Matthews, M.S. (Eds.). Mars. The University of Arizona Press, Tucson, USA. (***)
Melosh, H.J. 2011. Planetary Surface Processes. Cambridge Univ. Press, Cambridge, UK. (***)
Moore, P. 1998. On Mars. Cassel, London, UK. (*)
Murray, S. 2004. Eyewitnesses guide: Mars. Dorling Kindersley Book, London, UK. (*)
Mutch, T. A., Arvidson, R. E., Head III, J. W., Jones, K. L., & Saunders, R. S. 1976. The geology of Mars. Princeton University Press, Princeton, USA. (***)
Pyle, R. 2012. Destination Mars. New explorations of the red planet. Prometheus books, New York, USA. (*)
Sigurdsson, H., Houghton, B., Rymer, H., and Stix, J. (Eds.). Encyclopedia of Volcanoes Academic Press, New York, USA. (***)
Sparrow, G. 2014. Mars. A new view of the red planet. Quercus, London, UK. (*)
Taylor, F.W. 2010. The scientific exploration of Mars. Cambridge University Press, New York. (**)
Vogt, G.L. Landscapes of Mars. A visual tour. Springer Science+Business Media, USA. (*)
Zubrin, R. 2011. The case for Mars. The plan to settle the red planet and why we must. Free Press (Simon and Schuster), New York, USA. (*)

[2]This short list includes popular books (*), a few more advanced ones (**), and some more technical, reference volumes (***).

CPSIA information can be obtained
at www.ICGtesting.com
Printed in the USA
LVHW08s2126041018
592409LV00003B/7/P